Alkali-Aggregate Reaction and Structural Damage to Concrete

Alkali-Aggregate Reaction and Structural Damage to Concrete

Engineering assessment, repair
and management

Geoffrey E. Blight

Mark G. Alexander

CRC Press
Taylor & Francis Group
Boca Raton London New York

CRC Press is an imprint of the
Taylor & Francis Group, an **informa** business

A BALKEMA BOOK

CRC Press
Taylor & Francis Group
6000 Broken Sound Parkway NW, Suite 300
Boca Raton, FL 33487-2742

First issued in paperback 2018

CRC Press/Balkema is an imprint of the Taylor & Francis Group, an informa business

ISBN-13: 978-0-415-61353-8 (hbk)
ISBN-13: 978-1-138-07303-6 (pbk)

Typeset by Vikatan Publishing Solutions (P) Ltd, Chennai, India

Published by: CRC Press/Balkema
P.O. Box 447, 2300 AK Leiden, The Netherlands
e-mail: Pub.NL@taylorandfrancis.com
www.crcpress.com – www.taylorandfrancis.co.uk – www.balkema.nl

Library of Congress Cataloging-in-Publication Data
Applied for

**Visit the Taylor Francis Web site at
http:www.taylorandfrancis.com**

**and the CRC Press Web site at
http:www.crcpress.com**

Contents

Author biographies

GEOFFREY BLIGHT

Geoffrey Blight completed his Bachelor's and Master's degrees in Civil Engineering at the University of the Witwatersrand, Johannesburg and his PhD in Geotechnical Engineering at the Imperial College of Science and Technology, London, in 1961. The early years of his career were spent at the South African National Building Research Institute, Pretoria, where he was engaged in research on design, operation and safety of mine waste storage facilities, including waste rock dumps and hydraulic fill tailings storage facilities.

In 1969 Geoff Blight was appointed to the Chair of Construction Materials in the Department of Civil Engineering at Witwatersrand University. The field of study encompassed geotechnical engineering and concrete technology.

In 1978 he was commissioned to study and diagnose the cause of cracking occurring in a series of 15 to 17 year old reinforced concrete structures supporting the Johannesburg motorway system, and diagnosed the cause as AAR. Since then, he has researched and investigated several cases of deterioration by AAR, and has published widely on the subject. He and his co-author, Mark Alexander, spent a number of years in joint research on AAR and other aspects of the durability of concrete.

He was a corresponding member of the committee that produced the British Institution of Structural Engineers' guides on the structural effects of AAR published in 1989 and 1992 and, since 2002, has been a corresponding member of the RILEM Technical Committees TC 106 and TC 191 – ARP which have been investigating various aspects of AAR.

MARK ALEXANDER

Mark Alexander completed his Bachelor's, Master's and PhD degrees in Civil Engineering at the University of the Witwatersrand, Johannesburg and lectured in Construction Materials at Witwatersrand University for several years. In 1992 he was appointed to the Chair of Civil Engineering at the University of Cape Town where he has further developed his interests in concrete durability, including repair and rehabilitation of deteriorated concrete structures. He has published extensively and is active in international scientific circles. He co-authored the Book "Aggregates in Concrete" published by Taylor and Francis in 2005. He is currently Vice President of RILEM* and is scheduled to assume the RILEM Presidency in 2012.

---

* Reunion Internationale des Laboratoires et Experts des Materiaux.

Acknowledgements

The authors wish to salute the many engineers and scientists who have contributed to the understanding that we have of the AAR phenomenon in 2010, 70 years after it was first formally described by T.E. Stanton in the U.S.A. in 1940. We particularly acknowledge the great early contributions in the 1950's and 60's by Harold Vivian in Australia, Sydney Diamond in the U.S.A. and Gunnar Idorn in Denmark. In our own country, Bertie Oberholster led the field with his early investigations of the unexplained cracking of concrete structures in the Western and Eastern Cape Provinces. There are many other pioneers in the field that we have not mentioned, but we salute them all.

We also thank all the engineers and scientists who in later years, up to the present, have taken the trouble to share their experiences and ideas on combating and preventing the effects of AAR, thus adding to the pool of common knowledge of the subject. We particularly want to thank the co-authors of our papers on AAR. It was the efforts of all those mentioned above that made the writing of this book possible.

In Geoff Blight's case, the City Engineer and Deputy City Engineer of Johannesburg, Eric Hall and John Stewart, as well as the consulting civil engineer, Wally Schutte first involved him, in 1978, in this fascinating field. He thanks them for doing so.

Mark Alexander in his formative years as a young researcher, was most fortunate in having Geoff Blight, first as a supervisor and then as a colleague in the enabling environment at Witwatersrand University. No one could have wished for a more supportive and stimulating supervisor! He is very grateful for the opportunities then afforded him.

Both authors have had the good fortune to enjoy the support of dedicated and outstanding technicians. At the University of the Witwatersrand, Bob van der Merwe, Frans Wiid and Bob Anderson were indispensable to the success of our research. At both Witwatersrand and Cape Town, in addition to lab staff, many students have assisted from time to time in investigating various aspects of AAR. One of these was Yunus Ballim, who went on to become head of Civil Engineering and is now Deputy Vice-Chancellor of the University of the Witwatersrand. We are grateful to Yunus for his enthusiasm and support for writing this book.

We thank Alex Elvin, Professor of Civil Engineering at the University of the Witwatersrand, for contributing Section 4.8.6 on recent developments in instrumentation.

Maryanne Kelly prepared the diagrams and patiently made the seemingly endless changes, additions and deletions that were part of the process.

We both acknowledge and are extremely grateful for our families' unwavering support.

Geoff particularly thanks his wife Rhona for patiently typing yet another major manuscript from his almost illegible drafts.

Mark likewise acknowledges the patient and unswerving support of his wife Lyn over the years.

Unless otherwise acknowledged, all of the photographs were taken by Geoff Blight.

Geoff Blight
Johannesburg, September 2010

Mark Alexander
Cape Town, September 2010

List of mathematical symbols

ROMAN LETTERS

A	cross-sectional area [mm^2]
B, b	breadths [mm]
c	cohesion of concrete [MPa]
C	compressive force [kN]
or C	structural capacity [stress, force or moment units]
D	structural demand [stress, force or moment units]
D	diameter [mm]
E	elastic modulus [MPa, GPa]
f	direct stress (f_y = yield stress) [MPa, GPa]
$f(u)$ or $F(\ )$	function of quantity in ( )
H, h	heights [mm]
K	function of Poisson's ratio ν
L, l	lengths [mm]
P	load [kN, MN]
p^{II}	pore water suction [kPa]
P_f	probability of failure [dimensionless]
r	radius of water meniscus [mm or nm]
RH	relative humidity [dimensionless]
s_d	standard deviation [MPa]
T	tensile force [kN]
or T	surface tension of water [N/mm, kPam]
V	ultrasonic pulse velocity [m/s, km/s]
V_u	shear force at failure [N, kN]
W_f	failure load [kN]
$\bar{x}$	mean value

GREEK LETTERS

α	l.c.	alpha	ratio or coefficient [dimensionless]
β	l.c.	beta	ratio or coefficient [dimensionless]
γ	l.c.	gamma	shear strain [dimensionless]

ε	l.c.	epsilon	direct strain [dimensionless, mm/m = 1×10^{-3} = millistrain mm/km = 1×10^{-6} = microstrain]
η	l.c.	eta	viscosity [Pas, kNm/s]
μ	l.c.	mu	micro as in μm = micrometre or $\mu\varepsilon$ = microstrain
ν	l.c.	nu	Poisson's ratio [dimensionless]
π	l.c.	pi	ratio of perimeter to diameter of a circle [dimensionless]
ρ	l.c.	rho	density or unit mass [kg/m³ or g/cm³]
σ	l.c.	sigma	direct stress, strength or pressure [kPa = kN/m², MPa = N/mm²]
or σ			standard deviation of C or D
τ	l.c.	tau	shear stress or shear strength [kPa, MPa]
τ_u, τ_f			shear stress at failure
φ	l.c.	phi	angle of shearing resistance of concrete [degrees of arc]

Chapter 1

Alkali-aggregate reaction (AAR) and its effects on concrete – an overview

1.1   AAR AND ITS VISIBLE CHARACTERISTICS

The subject of the book is alkali-aggregate reaction and the structural damage it causes to concrete. The occurrence of AAR has been reported from every continent and both north and south hemispheres. It is a world-wide problem in concrete technology. The emphasis of the book is on engineering assessment, diagnosis and repair. There are many published conference and journal papers dealing with AAR, and the number of authoritative and specialized papers has increased substantially over the past three decades. For example, the number of papers presented to the Fifth International Conference on AAR in Concrete in 1981 was 32. At the 12th International Conference in 2004, 162 papers were presented. However, most of this literature is concerned with scientific rather than engineering aspects of AAR, i.e.: its chemistry, the nature of the various reactions, susceptible minerals, chemical kinetics, and factors influencing them, as well as the prediction of AAR-susceptibility. These scientific studies have been essential for our understanding of the AAR phenomenon and have assisted the cement and concrete industries to improve the performance of cement and aggregates and items made of concrete and to eliminate the use of marginal materials that may be AAR-susceptible.

The effects of AAR on the engineering performance of concrete artifacts and structures – whether of mass reinforced or prestressed concrete has been less well-covered in the engineering literature. At the 1981 AAR Conference, 4 of the 28 papers dealt with engineering aspects of AAR, while in 2004, 45 of the 162 papers were on engineering aspects. Hence not only has the number of engineering-orientated papers increased, but also their proportion (1 in 7 to 1 in 4.2). AAR has the primary effect of damaging concrete and causing some loss of performance in concrete structures, at the very least by the unsightly surface cracking it causes. The challenge to owners and operators of AAR-affected structures is how to assess these structures in terms of their engineering performance and safety, how to repair them and then how to manage the repaired structures. It must be recognised that, ultimately, civil or structural engineers have to assess, repair and manage the ongoing problems of AAR-affected structures. Owners of such structures will require reasonable assurance that their assets will retain their value and remain safe, functional and economic in terms of ongoing repair costs. Therefore, sound engineering knowledge of the mechanisms whereby concrete is damaged by AAR, its effects on the actual performance of concrete structures, as well as on assessment and repair techniques is needed.

Plates 1.1 to 1.7 show some examples and the appearance of damage to concrete structures, caused by AAR.

- Plate 1.1 shows the reinforced concrete foundation pier and Y-columns support-ing a road overpass in Quebec City, Canada. The white staining to the right of the view is caused by silica gel leached out of cracks. When photographed in 2000, the overpass had been in continuous use for more than 20 years and no repairs had been made.
- Plate 1.2 shows the AAR-damaged reinforced concrete abutment pier of a bridge over the Swan River, in Perth, Australia. This was completed in 1951 (Shayan and Lancucki, 1986) and cracking was noticed in 1966. When photographed in 1996, the abutment had received no attention in 45 years, except for superficial filling of the cracks in 1966, which failed to have any effect on the rate of crack-widening.
- Plate 1.3 shows AAR damage to an elevated freeway structure in Port Elizabeth, South Africa. The photograph shows the white stains of exuded gel. This struc-ture had not been repaired at all in its lifetime of over 40 years.
- Plate 1.4 shows AAR-damage to the wing-wall of a highway bridge in Johannes-burg, South Africa. Some cracking of the under-side of the bridge, to the right of the abutment is also visible. These cracks also have obviously not received any remedial attention, in this case, in 40 years.
- Plates 1.5 and 1.6 are particularly interesting. Plate 1.5 is a view of the cen-tral abutment of a double arch railway bridge in Johannesburg, photographed in 1978. This shows all the symptoms of apparently quite severe AAR-attack, including cracking of the concrete and exuding gel. The bridge received no reme-dial attention and looked very much the same as this until 2010 when the city was cleaned up prior to hosting the world cup soccer tournament. The cracks were filled, the drainage of water from the fill above the arch ribs was improved and the bridge was given a coat of elastomeric paint. A few months later, the filled cracks were still visible, and some of the more severe cracks, which coincided with construction joints, were re-opening. The thirty two years (at least) of serv-ice in a damaged condition, shows that AAR-attack certainly need not terminate the useful life of a structure. Although damage can temporarily be hidden by a cosmetic treatment, it cannot be completely cured with cosmetics alone.
- Finally, Plate 1.7 shows AAR cracking in the concrete base block of an overhead lighting pylon. This shows the exudation of both white gel and brown rust stain-ing, from reinforcing in the block.

If this pictorial overview leads the reader to conclude that, in many cases, AAR damage is not as serious as it appears, the reader would be correct. Nevertheless, there are some cases in which prompt remedial action appears to be called for, and some of these will be explored in Chapter 5. However, no case is on record, to the authors' knowledge, when structural collapse has had AAR as its primary cause.

This book will cover the basic science of AAR, because an understanding of the phenomenon is essential for a better appreciation of its engineering effects and consequences. However, it will not provide a comprehensive overview of this area. Other texts should be consulted for more scientific detail if needed. The book seeks to provide a useful ready-reference source with many illustrative case histories, for practicing structural and materials engineers who are faced with the issues of assessing the performance and safety of AAR-affected concrete structures as well as planning and executing repairs and post-repair management.

1.2 THE CHEMICAL CHARACTERISTICS OF AAR

Alkali-aggregate reaction occurs as a chemical reaction between alkalis in the concrete matrix and certain minerals in the aggregates which are susceptible to attack by strong alkalis. The phenomenon was first observed in the USA in the 1920s and was formally identified in the engineering literature by Stanton (1941). It became a major focus of attention, both in terms of public awareness (the so-called 'concrete cancer' scare in the UK in the 1980s), and research effort and expenditure, in various parts of the world from the 1970s onwards. To be initiated and to proceed, AAR requires the presence of sufficient moisture. As shown by Plates 1.1 to 1.7, it may result in substantial damage to concrete structures, in the form of both micro- and macro-cracking within the matrix and the aggregates, as well as loss of bond to reinforcing. Externally, these effects appear as unsightly and sometimes alarming surface cracking and staining, with crack widths that may exceed 10 mm. Cracking usually becomes evident several (5 to 15) years after construction and the reaction may then progress continuously or intermittently for many years, or to all intents and purposes, cease completely.

Provided there is sufficient moisture, the reaction usually forms a gel which expands within pores and microcracks in the concrete. This causes characteristic map cracking and in addition, cracking very often occurs parallel to the direction of the dominant stress system or main reinforcing. For example, compressive stresses in a column attacked by AAR may result in longitudinal cracking. AAR may also be identified by gel weeping from cracks and white staining below cracks. (See Plates 1.3, 1.5 and 1.7.) During prolonged dry weather the gel becomes desiccated and appears as a whitish powder covering crack surfaces when these are exposed. (An exception to this rule is alkali-carbonate rock reaction). Plate 1.7 shows typical white staining and cracking in the mass concrete base of an overhead lighting pylon as well as brown staining from corroding holding-down bolts.

1.3 GUARDING AGAINST AAR

Concrete, as with any other construction material, deteriorates with time. In the vast majority of cases, this deterioration is sufficiently slow so that the structure in which the concrete is used has an extremely long life expectancy. Ancient Roman structures, made of an early form of concrete are witness to this fact. As an outstanding example, the dome of the Pantheon in Rome spans 43 m and also rises 43 m (King, 2000). It was constructed between AD 118 and 128 and contains 5000 tonnes of pozzolan-bound concrete that used low-density pumice aggregate. As a true dome, it remained unequaled in span for over 1300 years. Plate 1.8 shows an engraving of the interior of the Pantheon in which the circular opening or oculus at the crown of the dome, designed both for illumination and to reduce weight, can be seen. Many modern concrete structures, however, deteriorate at an unacceptable rate. The deterioration usually results from a combination of extrinsic and intrinsic factors, such as the inter-action of the external environment with the material itself as in the case of sulphate attack on concrete. Figure 1.1 gives an overview of concrete deterioration mecha-nisms in the context of durability, and locates the AAR phenomenon as an internal chemical instability, influenced by the concrete constituents (cements and aggregates) and external environment (temperature and moisture), i.e. AAR combines extrinsic and intrinsic factors.

The particular concern of this book is the deterioration of concrete caused by the attack on aggregate minerals by alkalis contained in the concrete. The aggregates in concrete are usually harder than the matrix or binder phase, and therefore tend to be more durable. However, there are instances when the aggregates themselves contribute directly to concrete deterioration or are susceptible to external aggressive environments.. These processes usually require external water to be available, and AAR is typical of this form of deterioration. It is difficult to separate extrinsic and intrinsic mechanisms. For example, the water that maintains the AAR may enter the concrete from external sources, but the reaction is intrinsic to the composition of the concrete.

Engineers may appear to take aggregates and their effects on concrete seriously only when they present a problem, such as AAR. This reflects a widespread view that concrete aggregates are inert fillers, with little influence on the properties of the

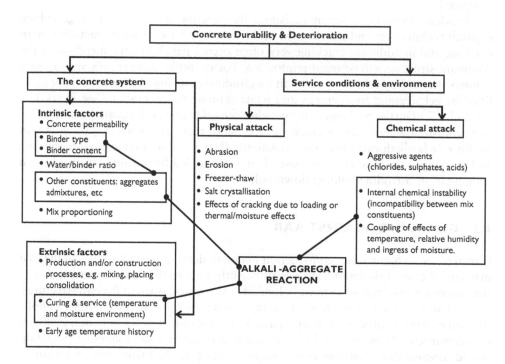

Figure 1.1 Concrete and environment: factors influencing the durability of concrete (After Oberholster, 2001).

concrete. However, aggregates should always be viewed as integral and important constituents of concrete as they typically make up between 65% and 80% of the volume of the concrete. In the context of AAR, engineers should appreciate that certain aggregates can give rise to damaging and detrimental effects in concrete, which can be either physical or chemical in nature or have combined physical and chemical effects. Particularly if new or untested aggregate sources are used, engineers should know how to identify or have possible alkali-susceptible aggregates identified and whether or not to permit their use in concrete so that the material and the structure from which it is made are not adversely affected.

1.4 MAIN TYPES OF AAR AND THE APPEARANCE OF FRACTURES CAUSED BY AAR

Three main types of AAR have been identified, depending on the aggregates: Alkali-Silica Reaction (ASR), Alkali-Silicate Reaction, and Alkali-Carbonate Rock Reaction (ACR).

1.4.1 Alkali-silica reaction (ASR)

This is the most common form of AAR, and involves reactions between alkaline pore solution in the concrete and certain forms of silica such as those found in volcanic glasses, cristobalite, tridymite, and opal. It also includes reactions with aggregates containing chert, chalcedony, microcrystalline quartz, cryptocrystalline quartz, or strained quartz. The distinguishing features are reactive forms of silica, amorphous or altered, that are not chemically stable, i.e. they are 'alkali-susceptible'. Typical rocks in which these occur are greywacke, quartzite, hornfels, phyllite, argillite, granite, granite-gneiss, and granodiorite. ASR is associated with the formation of expansive alkali-silica gel in concrete. The governing equation for the reaction is given below

$$2(\text{Na/K})\text{OH} \quad + \quad \text{SiO}_2 \quad + \quad \text{H}_2\text{O} \quad \rightarrow \quad \text{Na}_2\text{SiO}_3 \cdot 2\text{H}_2\text{O}$$
$$\text{Alkali} \qquad\qquad \text{Silica} \qquad \text{Water} \qquad \text{Alkali-silica gel} \tag{1.1}$$

The degree of reactivity of the silica minerals in the aggregate depends mainly on their crystal structure. Highly disordered, semi-amorphous structures such as opal are very reactive, while ordered structures such as unstrained quartz are normally unreactive. (Details of the varieties of silica and their reactivities are given in Chapter 2 (Table 2.2).

A so-called pessimum (i.e. worst) condition may be associated with ASR. This is the amount of aggregate or alkali at which expansion is a maximum, i.e. where the combination of reactive silica and alkali is such that the greatest expansion occurs. For aggregate and alkali combinations other than the pessimum, potential expansions will be less.

The distinctive microstructural features of ASR that can be identified in broken or cracked lumps of concrete are usually a whitish reaction product (desiccated gel) often with the voids in the concrete filled with this product, reaction rims around aggregate particles, cracks through aggregates which are also sometimes filled with gel, matrix cracks which are often contiguous with the aggregate cracks, and loss of bond between aggregate and matrix. ASR cracks in aggregates often derive from incipient or pre-existing microcracks that are penetrated by alkaline pore solution. Crushed aggregates may have abundant incipient fractures and therefore this feature is often noted in crushed aggregates.

Plates 1.9 and 1.10 illustrate some of these features. The photographs show the polished surfaces of two cores of AAR-damaged concrete. Features are:

- Plate 1.9 shows darkened reaction rims around three large aggregate particles. The white outer rims of the particles show that the aggregate has lost its bond with the surrounding mortar.
- Plate 1.10 shows (to a slightly larger magnification) a large aggregate particle with a dark reaction rim, cracks within the particle and a crack around the perimeter.

In both cases, the scale shown is in mm.

1.4.2 Alkali-silicate reaction

According to Oberholster (2001), the term alkali-silicate reaction was originally introduced to differentiate AAR involving reactions with aggregates from rocks such as phyllites, argillites and certain greywackes which contain phyllosilicates (e.g. vermiculite, chlorite and mica) from AAR involving forms of silica. In this case, the reaction is with the silica in the combined form of the phyllosilicates, rather than free silica. These reactions are complicated and difficult to characterise but may be expansive. If the expansion with these rocks is caused by the presence of finely divided silica in the rocks, it is more strictly an alkali-silica reaction. In any case, there is debate as to whether the basic reaction is different from conventional alkali-silica reaction.

1.4.3 Alkali-carbonate rock reaction (ACR)

Alkali-carbonate reaction does not produce a gel. Instead, coarse aggregate particles expand due to alkali hydroxide reaction with small dolomite crystals in a clay matrix, resulting in a dedolomitization reaction (i.e. the breakdown of dolomite into brucite and alkali carbonate, i.e. calcite – see equation (1.2).

$$2(Na/K)OH + CaMg(CO_3)_2 \rightarrow CaCO_3 + Mg(OH)_2 + (Na/K)_2CO_3 \quad (1.2)$$

Alkali   Dolomite   Calcite   Brucite   Alkali carbonate

This type of AAR is limited to carbonate aggregate containing clay and dolomite, such as certain argillaceous dolomitic limestones. Extensive cracking can result (Swenson and Gillott, 1964, Dolar-Mantuani, 1983), but expansion seems to depend on the nature of the minerals and the micro-textures. The most common mechanism involves dedolomitization of the minerals by alkalis followed by swelling of the interstitial clay caused by increasing moisture content. The reaction does not occur with normal calcitic limestones.

Alkali-carbonate rock reaction is not widespread and is encountered mainly in Canada where alkali susceptible carbonate rocks occur in the Gull River Formation along the southern margin of the Canadian Shield from Midland to Kingston in Southern Ontario. The same reactive rocks occur in the Ottawa-St Lawrence region. While ACR is locally very important, it will not be further covered in this book due to its limited occurrence. This book will deal in the main with the much more common phenomenon of alkali-silica reaction.

1.5   CHEMICAL MECHANISMS OF AAR

Helmuth and Stark (1992) provided an explanation for the ASR mechanism, reviewed by Mindess, et al. (2003) as follows:

Helmuth and Stark observed that the alkali-silica reaction results in the production of two-component gels – one component is a non-swelling calcium-alkali-silicate-hydrate (C-N(K)-S-H) and the other is a swelling alkali-silicate-hydrate (N(K)-S-H). When the alkali-silicate reaction occurs in concrete, some non-swelling C-N(K)-S-H is always formed. The reaction will be safe if this is the only reaction product, but unsafe if both gels form. The key factor appears to be the relative amounts of alkali and reactive silica. The overall process proceeds in a series of overlapping steps:

- In the presence of a pore solution consisting of H_2O and Na^+, K^+, Ca^{2+}, OH^- and $H_3SiO_4^-$ ions (the latter from dissolved silica), the reactive silica undergoes depolymerization, dissolution, and swelling. The swelling can cause damage to the concrete, but the most significant volume change results from cracking caused by subsequent expansion of reaction products.
- The alkali and calcium ions diffuse into the swollen aggregate resulting in the formation of a non-swelling C-N(K)-S-H gel, which can be considered as C-S-H containing some alkali. The calcium content depends on the alkali concentration, since the solubility of CH is inversely proportional to the alkali concentration.
- The pore solution diffuses through the rather porous layer of this C-N(K)-S-H gel to the silica. Depending on the relative concentration of alkali and the rate of diffusion, the result can be safe or unsafe. If CaO constitutes 53 per cent or more of the C-N(K)-S-H on an anhydrous (without water) weight basis of the gel, only a non-swelling gel will form. For high-alkali concentrations, however, the solubility of CH is depressed, resulting in the formation of some swelling N(K)-S-H gel that contains little or no calcium. The N(K)-S-H gel by itself has a very low viscosity and could easily diffuse away from the aggregate. However, the presence of the C-N(K)-S-H results in the formation of a composite gel with greatly increased viscosity and decreased porosity.
- The N(K)-S-H gel attracts water due to osmosis, which results in an increase in volume, local tensile stresses in the concrete, and eventual cracking. Over time, the cracks fill with reaction product, which gradually disperses under pressure from the point of its initial formation.

Mindess, et al. (2003) point out that the higher the alkalinity of the cement (i.e. the OH⁻ concentration of the pore solution), the greater the solubility and dissolution rate of amorphous silica. The rate and extent of the first step above depends on aggregate porosity; this also governs whether alkali attack takes place throughout the particle or initially only on the surface. Further, the formation of a swelling gel depends on the relative amounts of silica and alkali and the degree to which the alkali is 'tied' by the silica, causing a decrease in the pH of the pore solution. This also explains the occurrence of a "pessimum percentage" of alkali.

Compressive stresses applied to concrete can reduce or slow expansion due to alkali-silica reaction. Alkalis in the form of salts can accelerate the alkali-silica reaction, and thus seawater can exacerbate the situation. However, pozzolans in the form of reactive silicas can reduce the severity of the reaction if used in adequate amounts. This seems to be due to the fact that finely divided silica (<0.15 mm) encourages a rapid reaction without deleterious effects, with the reaction moving rapidly through the fourth stage listed above, the reaction products, being well distributed, are little affected by gel viscosity and a uniform distribution of reaction products results. This decreases local concentration gradients and subsequent osmotic pressures. The action of silica fume as well as fly ash in eliminating harmful effects of ASR is an example of this effect.

1.6   NECESSARY AND SUFFICIENT REQUIREMENTS FOR AAR TO OCCUR

There are three requirements necessary for AAR to occur: a source of alkalis, a source of reactive silica in the aggregate, particularly the coarse aggregate, and an environment contributing sufficient moisture to cause swelling of the gel.

1.6.1   Alkalis

The alkalis must be present in the pore solution of the concrete, and the primary source of alkalis is the cement or binder itself. This contains metal alkalis – sodium and potassium hydroxide – as well as liberal amounts of calcium hydroxide. In AAR, cement alkalis are quantified by the **equivalent sodium oxide**, expressed as a percentage by mass of the cement, as follows

$$\%Na_2O_{eq} = \%Na_2O + 0.658\%K_2O \tag{1.3}$$

The constant 0.658 in Eqn. 1.3 (often rounded to 0.6) is from the ratio of the atomic mass of Na_2O to K_2O, and permits the **equivalent** effect of potassium in contributing alkalis to be quantified.

Other sources of alkali in concrete can be the environment, such as marine or de-icing salts – which can be particularly aggressive in respect of AAR – or chemical admixtures usually containing sodium. Alkalis sometimes also derive from the aggregates themselves. For examples, alkali-containing minerals such as feldspars may react with the calcium hydroxide released by cement hydration. Aggregates containing artificial glass, and volcanic glass present in some rhyolites, basalts, and andesites may also release alkalis. (Stark, 1978, Stark and Bhatty, 1985). The alkali contribution from aggregates, though difficult to assess, may be important enough to take into account.

Portland cements may be classified as high, medium, or low alkali, with cements having Na_2O_{eq} greater than 0.6 per cent being regarded as high alkali cements. However, it is the **total** alkali content in the concrete that is more important than the alkali content of the cement itself. The total alkali content is determined from the alkali content of the cement, the cement content, and the proportion of alkalis released during hydration and available for reaction (termed 'active alkalis'). This latter factor varies for different cements but ranges typically between 70–100 per cent of total equivalent Na_2O, with a typical value for most portland cement clinkers of 85 per cent. Practically, the total alkali content of the concrete is calculated as the product of the cement alkali content and the cement content of the mix. This value must be limited depending on the type of reactive aggregate. Figure 1.2 shows the increasing expansions of concretes made with the same alkali-susceptible aggregate and increasing Na_2O equivalents.

AAR-deteriorated concrete is often found to have an alkali content that far exceeds that of the original concrete mix. It has been suggested, e.g. Diamond (1981) that the high alkali values result from concentration by leaching of alkali and that leaching of alkali by intermittent wetness explains seasonal periodic expansion (well illustrated by Figure 3.1b). The authors (Blight, et al., 1984) have been unable to find any evidence of appreciable alkali migration by leaching. In one case a column supporting

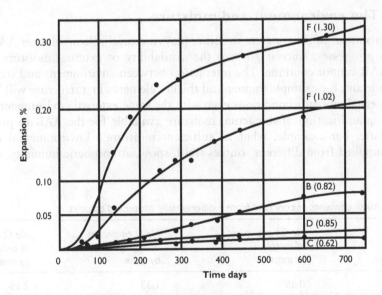

Figure 1.2 Effect of cement alkalinity on ASR. Malmesbury coarse aggregate (greywacke) in combination with different cements. Cement content of concrete = 350 kg/m³. Stored under ASTM C227 conditions above water in sealed containers at 30°C. The figures in brackets refer to the total %Na_2O equivalent of the cement (Oberholster, 1986).

the west side of a motorway structure was cored in order to establish if there was any differential concentration of alkali. The north and west sides of the column are exposed to the weather and therefore subject to periodic wetting and drying, whereas the south and east sides are protected from rain and sun by the structure overhead. The results of the analysis of the cores are shown in Table 1.1. There appears to be no preferential concentration of alkali in the column.

Cores were also taken from the concrete walls of two water reservoirs where the concrete is subject to continual leaching by outward seepage of water. The alkali analysis of these two sets of cores is shown in Figure 1.3. There appears to be a very slight trend for alkali contents to increase towards the outer face of the concrete, but the trend is by no means clear cut.

The reservoirs had been in service for 20 years, and the cores were taken from close to the bottom of the circular reservoir walls, where the depth of retained water was 10 m. For the 300 mm thick wall of reservoir 1 the seepage gradient was therefore 100/0.3 = 333 kPa/m. For the 400 mm thick wall of reservoir 2 the seepage gradient was 250 kPa/m. Hence there must have been sufficient saturated seepage flow to cause leaching of unbound salts.

1.6.2   Reactive silica

ASR is governed by the nature and quantity of reactive silica present. Chapter 2 contains details of the rocks and minerals that are susceptible to alkali attack (Table 2.2), and should be referred to for information.

1.6.3   The environment and moisture

The environment of the concrete structure plays a crucial role in whether AAR develops and progresses, since it governs the availability of external moisture without which AAR cannot continue. The interaction between environment and structure is also important, for example thinner and thicker elements in a structure will have different internal moisture conditions even with the same external environment.

The quantification of the actual moisture available for the AAR to progress is problematic. For example, what is 'sufficient moisture'? Environmental moisture can be supplied from different sources (rain, snow, atmospheric humidity, moisture

Table 1.1 Alkali contents of concrete from a differentially protected column.

Sample from	Na_2O^* equivalent of mortar % by mass	Na_2O equivalent of aggregate % by mass	Na_2O equivalent of cement % by mass
West	0.86	1.33	2.15
North	0.56	0.72	1.40
South	0.80	1.35	2.00
East	0.70	1.26	1.75

* Equation (1.3).

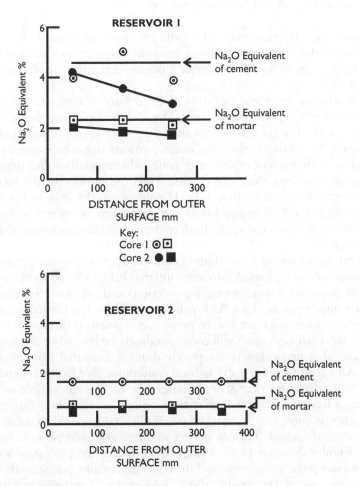

Figure 1.3 Results of investigation into possible leaching of alkalis to the surface of concrete water reservoir walls.

from the underlying soil, etc.) and how this moisture interacts with the structure is critical. In drier climates where environmental moisture is limited, moisture penetrating the structure may derive, not directly from the surroundings, but from sources such as poor drainage (e.g. leaking bridge joints or ponding on the structure), or from condensation. Such sources may be intermittent but may result in a reservoir of moisture that builds up in the structure, and causes ongoing AAR. Thus, both when designing a structure in a situation where AAR may be a risk, or when diagnosing a structure in which AAR is suspected, a clear picture of potential or available moisture sources must be obtained. This is also very often the key to remediation – if the moisture source is not properly identified, the remediation is unlikely to be successful. (For more details, see Section 3.2.)

Other factors that should be considered are:

- Environmental effects, specifically moisture, may be greater in thinner elements where the interaction between external moisture and internal conditions is greater. On the other hand, thinner elements dry more rapidly, and this may retard AAR.
- There is usually sufficient moisture 'locked into' a concrete structure from the concrete mixing water to trigger ASR, other conditions being met. Thus, in larger or thicker members even in drying environments, internal moisture will usually be retained while the outer portions dry. The expansive reaction occurs internally but not in the outer zones which usually suffer drying shrinkage. Surface cracking therefore results, while the interior of the member may remain uncracked (see Section 3.2). The major cracks seen in Plates 1.5 and 1.7 resulted from AAR expansion of the internal or heartcrete while the outer covercrete went into tension due both to drying and the restrained expansion of the heartcrete.
- In general, structures in moist, humid climates appear more susceptible to AAR than those in drier regions. A minimum internal R.H. of 95 per cent has been suggested as necessary to permit ongoing reactions and expansion (see Section 5.2). Moisture must be present for AAR gel to form. However, for subsequent expansion to occur moisture must also be present, and clearly if this moisture is limited or excluded, then expansion will correspondingly be limited or negligible.
- Walls and abutments that retain poorly drained saturated fills can also suffer from AAR even in relatively dry natural conditions. (See Plates 1.4 and 1.5.)
- Concrete that remains permanently damp appears less susceptible to AAR. For instance, it has been observed many times (Blight, et al., 1981) that road bridge and motorway piers may be cracked where they are exposed to the atmosphere, and particularly where they are subject to differential and periodic wetting as a result of faulty drainage of the bridge deck. Where the piers pass into the soil under surface paving, however, and the concrete remains permanently damp, the cracking disappears. Presumably, this is because the covercrete is not subjected to shrinkage, but expands at the same rate as the heartcrete. AAR damage has never been observed by the authors to have affected buried pile caps or piles, again presumably because no shrinkage had occurred in the covercrete. This evidence has been contradicted by reports from the United Kingdom by Wood and Wickens (1983) who, in describing AAR attack on a car park structure, state: "The ... remedial work necessitated the opening up of the foundations. They were found to be severely cracked. The vulnerability of buried foundations to attack because of their damp environment has also been noted on other structures". Hence the fact that permanently damp concrete has not been observed by the authors to be subject to AAR apparently does not preclude the possibility of some damage occurring, but if AAR-expansion occurs in uniformly wet concrete, it may not cause cracking.
- There is experimental evidence that concretes that are mechanically microcracked by loading (which is the 'normal' design assumption for reinforced concrete members, at least in highly stressed zones) experience enhanced rates of AAR and consequent damage. In such cases, a dormant stage occurs during which the

pre-existing microcracks are filled by gel, followed by active AAR expansion leading to additional cracking.

1.7 WHAT IS STILL TO COME

This introductory chapter is followed by five chapters dealing, in Chapter 2, with:

* Inspections, assessment and diagnostic investigations of structural damage to concrete. If, during a preliminary inspection, AAR is suspected as a possible cause of the damage, the inspection would be followed by in situ and laboratory diagnostic tests, including petrographic examinations and, if considered necessary, chemical and physical testing. The susceptibility of aggregates to AAR is also considered in Chapter 2.
* Chapter 3 deals with the effects of AAR on the engineering properties of concrete, as revealed by laboratory tests on cores taken from AAR-damaged structures. The emphasis is on the engineering properties of cores that have suffered AAR-damage under natural conditions. The differences in the behaviour of cores taken from real structures, and that of specimens of concrete made in a laboratory and subjected to artificially accelerated AAR, are sufficiently large to lead the authors to believe that the results of laboratory-accelerated tests may be qualitatively useful, but are often quantitatively misleading.

    The chapter includes brief descriptions and analyses of a number of destructive laboratory tests on cores that are used to assess various aspects of the strength and elastic properties of concrete, as well as the creep under load of AAR-damaged concrete, the effects of compression on expansion caused by AAR and the compressive restraint exerted by reinforcing.

    Fractures in reinforcing found in AAR-affected concrete have recently been reported, especially in Japan, and the causes and prevention of such damage is also examined.
* Chapter 4 considers the assessment of the risk of structural failure as a result of AAR-damage. Two approaches are taken. Part 1 of the chapter examines the statistical calculation of the probability that failure will occur. This is the type of analysis that would be undertaken in assessing the condition of a damaged structure, and as a tool in deciding on an appropriate course of action. Action considered necessary could range from "do nothing" to cosmetic repair, to various degrees of structural repair, to partial or complete demolition and re-building. The authors hasten to reiterate that very few cases of partial demolition and rebuilding are on record and none of complete demolition. As far as the authors know, no structure has ever collapsed solely as a result of AAR-damage, but the possibility exists, and must be guarded against.

    Part 2 deals with full-scale test loading of complete structures or parts of complete structures, as a means of assessing the risk of failure. The chapter includes descriptions of the instruments commonly used in full-scale loading tests to measure deflection or sway, rotation, strain and temperature and looks at recent developments in the field of instrumentation. Non-destructive testing by means of ultra-sonic pulse velocity measurement is also described and its utility is assessed. This is followed by sections describing the planning, preparation and performance of an in situ loading proof test on a full-scale structure.

The remainder of Part 2 describes six special or once-off test loadings that have been done in various parts of the world. These are full-scale proof loading tests on two double cantilever supports for overhead motorways in Japan, a motorway portal frame in South Africa, a motorway bridge in Denmark, a concrete road pavement, an underground mass concrete air receiver plug and an industrial structural slab in South Africa. The criteria and purpose for their choice were to include as large a variety of structural types and climatic conditions as possible, and to be carefully planned, well instrumented tests with adequate analysis and good reporting. These are followed by case histories of routine, periodic test loadings on a loading jetty in Australia and a series of road bridges in north eastern France, and tests on relatively small structural components. These were components or parts removed from structures damaged by AAR, and tested in laboratories. They include prestressed concrete railway sleepers from a dedicated ore export rail line in South Africa, beams sawn from the decks of two viaducts in the Netherlands and prestressed concrete planks from the deck of a road bridge in Australia.

- Chapter 5 examines the problems of repair and rehabilitation of AAR-damaged structures by several possible methods. Criteria for inclusion in this section were similar to those that applied to the previous section. The repair method had to be well documented and the repaired structure has to have been monitored by observation, and preferably by measurement, for several years after repairing, to give assurance that the repair was successful. The types of repair reported in chapter 5 are:
  - Arresting the AAR process and rendering it dormant. Descriptions of projects to investigate this approach in Iceland, France and South Africa are given. Both of the later investigations included laboratory phases, followed by field tests on full-scale structures.
  - Sealing the cracks and repairing the damaged concrete by resin injection. Only one successful, well documented and monitored case was found. This is a sports grandstand in South Africa that, after being severely damaged by AAR, was restored to full functionality and an excellent appearance by resin injection and a surface treatment. This example has been monitored since 1992.
  - Repair by externally applied stressing. Two examples from South Africa are described. These were repaired in the early 1980's, and 30 years later, with no additional attention, still appear in an excellent state. In a third example, from Japan, an AAR-damaged bridge pier was strengthened by encasing it in a horizontally stressed concrete encasement. This example has been monitored for at least 6 years and has proved very successful.
  - Repair by partial demolition and reconstruction. Chapter 5 deals with three cases: the rehabilitation of a series of railway bridge piers in Canada, refurbishment of a rail bridge underpass, also in Canada and repair of a double decked motorway portal frame in South Africa. In all cases the operations were successful.

- A number of cases dealing with dimensional changes in underground hydro-electric power stations by means of slot-cutting and excision of swollen concrete, as well as the effects of AAR in large mass concrete or double-curvature arch dams are described. The cases are drawn from Brazil, Canada, South Africa and the U.S.A.
- The rehabilitation of AAR damage to an unreinforced concrete highway pavement is also described.
- The repair of broken reinforcing discovered in AAR-damaged bridge structures in Japan is described. This is the only repair described that is likely to prove ineffective, because the basic cause of the problem was not recognized.
- Finally, as an epilogue, a check-list is given of important structural consequences of AAR.

REFERENCES

Blight, GE, McIver, JR, Schutte, WK & Rimmer, R 1981, 'The effects of alkali-aggregate reaction on reinforced concrete structures made with Witwatersrand quartzite aggregate', *5th Int. Conf. on AAR in Concrete*. Cape Town, South Africa. Paper S252/15.

Blight, GE, Alexander, MG, Schutte, WK & Ralph, TK 1984, 'The repair of reinforced concrete structures affected by alkali-aggregate reaction', *International Concrete Symposium, Concrete Society of Southern Africa*, Halfway House, South Africa, pp. 158–190.

Diamond, S, et al., 1981, *Discussion of Blight, et al., 1981*.

Dolar-Mantuani, L, 1983, *Hardbook of concrete aggregates*, Noyes, New Jersey, U.S.A.

Helmuth, R & Stark, D 1992, 'Alkali-silica reactivity mechanisms', in *Materials science of concrete III*, ed JP Skalny, American Ceramic Society, Westerville, U.S.A., pp. 131–208.

King, R 2000, *Brunelleschi's Dome*, Penguin, New York, U.S.A.

Mindess, S, Young, JF & Darwin, D 2003, *Concrete*. Practice-Hall, New Jersey, U.S.A.

Oberholster, RE 1986, 'Alkali-aggregate reaction', in *Fulton's concrete technology*, eds BJ Addis & G. Owens, 6th edn, Cement and Concrete Institute, Midrand, South Africa.

Oberholster, RE 2001, 'Alkali-silica reaction', in *Fulton's concrete technology*, eds BJ Addis & G Owens, 8th edn, Cement and Concrete Institute, Midrand, South Africa.

Shayan, A & Lancucki, CJ 1986, *7th Int. Conf. on AAR in Concrete*, Ottawa, pp. 392–402.

Stanton, TE 1940, 'The expansion of concrete through reaction between cement and aggregate'. *Proc. American Society of Civil Engineers*, vol. 66, pp. 423–431.

Stark, D 1978, 'Alkali-silica reactivity in the Rocky Mountain region', *4th Int. Conf. on AAR in Concrete*, Indiana, U.S.A., pp. 235–243.

Stark, D & Bhatty, MSY 1985, 'Alkali-silica reactivity: effect of alkali in aggregate on expansion', in *Alkalis in concrete*, ASTM STP 930, Pennsylvania, U.S.A., pp. 16–30.

Svenson, GE & Gillot, JE 1964, *Alkali-carbonate rock reaction*, Highway Research Record No. 45, Highway Research Board, Washington, D.C. (U.S.A.) pp. 21–40.

Wood, JGM & Wickens, PJ 1983, *6th Int. Conf. on AAR in Concrete*, Copenhagen, Denmark, pp. 176–185.

PLATES

Plate 1.1 Multiple cracks on surface of road over-pass pier.

Plate 1.2 Cracking on abutment of bridge over river.

Plate 1.3 Cracking in elevated freeway structure.

Plate 1.4 Cracking in wing-wall of bridge abutment.

Plate 1.5 Poor condition of railway arch bridge photographed in 1978. (view from right)

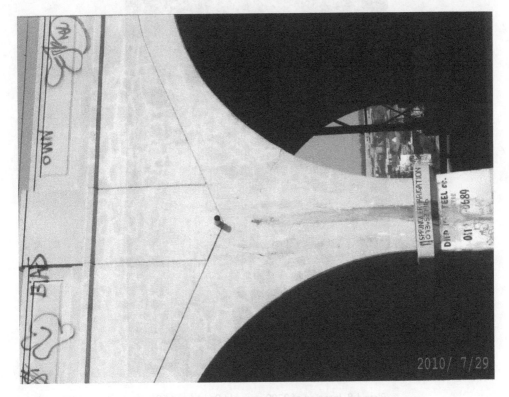

Plate 1.6 Condition of the same bridge shown in Plate 1.5, in 2010. (view from right)

Plate 1.7 Gel and rust staining of foundation of overhead lighting tower. (view from right)

Plate 1.8 Interior of 2200 year-old Pantheon in Rome.

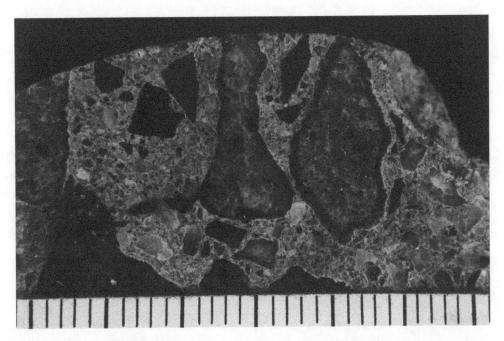

Plate 1.9 Polished surface of a concrete core showing reaction rims and indications of loss of bond with mortar. Scale is mm.

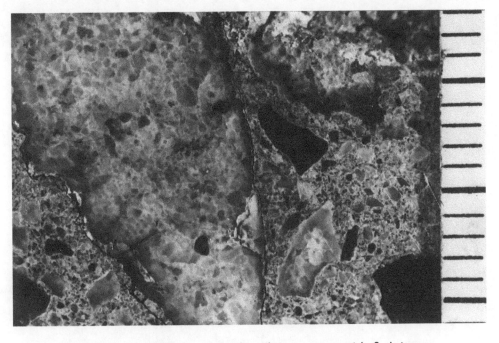

Plate 1.10 Similar to 1.9, showing cracks within aggregate particle. Scale is mm.

Chapter 2

Diagnostic investigations and tests and their interpretation

2.1 INVESTIGATION OF THE CAUSE OF CRACKING IN A CONCRETE STRUCTURE

There are many reasons why a concrete structure may crack. It is, however, only rarely that AAR will be the cause of cracking in building, retaining or bridge structures. If founded on soft to firm clay or on loose to medium dense silt or sand, differential settlement is often the cause of structural cracking. Settlement may also cause cracking if the structure is founded on poorly, or unevenly compacted fill, or partly in cut and partly on fill. In semi-arid to arid climatic zones, cracking may result from differential expansion (or heave) of desiccated clays or clayey soils. Another cause of cracking could be differential shrinkage caused by the abstraction of water from the foundation soil by nearby vegetation, or swelling by the local addition of water by irrigation of nearby vegetation. In seismically active areas of the world, cracking may originate from seismic accelerations. A similar range of sources of movement could cause cracking in soil or water retaining structures, or in concrete structural pavements. Cracking could also be the result of plastic and/or drying shrinkage of the concrete, thermal contraction or frost attack.

Apart from these purely physical causes of cracking, there is a variety of possible chemical causes, such as sulphate attack, or corrosion of the reinforcing as a result of chloride ingress. Attack by AAR could be grouped with chemical causes. Finally, the loading for which the structure was designed may have been inadequate, or the structural detailing may have been incorrect, or knowledge of the properties of the structural materials could have been lacking. Unfortunately, these three causes occur the most often, and appear to have led to more disastrous structural collapses than any other cause. A count of the causes of the failures described in the American Society of Civil Engineer's book "Failures in Civil Engineering" (Shepherd and Frost, eds., 1995) showed that of the 40 failures that occurred in the 20th century, described in the book, 30 were the result of under-estimation of loading, lack of knowledge of material capabilities or incorrect structural detailing. Of the 8 disastrous failures in reinforced concrete structures, 6 were the result of these three causes.

What follows is based on Larbi, et al. (2004):

A careful examination must be made of any concrete structure that has cracked, in order to diagnose the true cause or causes of the cracking. Figure 2.1 shows a scheme for such a preliminary diagnostic examination.

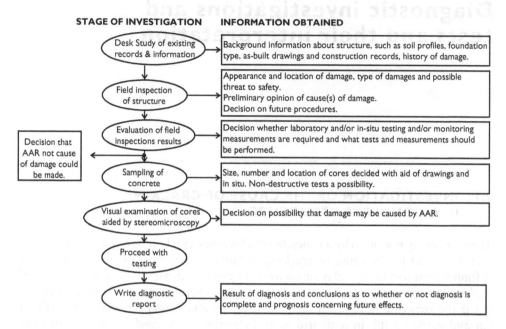

STAGE OF INVESTIGATION   INFORMATION OBTAINED

Figure 2.1 Scheme for preliminary diagnostic examination of cracked structure to ascertain probable cause of cracking.

2.1.1   Planning the site inspection

This requires a review of all available documents and information relating to the structure, prior to the inspection. Information should be sought on:

- the type of structure,
- environmental conditions to which it is or has been exposed,
- date of completion,
- details of any prior major or minor additions or remedial work,
- design loading and construction details,
- foundations and soil profiles,
- drainage system,
- joints,
- position, quantities and arrangement of reinforcement,
- details of concrete, including mix design and sources of cement and aggregates,
- reports of previous inspections or testing, either routine or specialized,
- dates on which signs of damage or deterioration were first noticed,
- any other information that may be relevant for the field inspection and the investigation, such as the performance or possible problems encountered with other nearby structures of similar concrete construction and age, and
- any changes made to the structure from the original approved design.
- "as built" drawings are very often not as actually built and must be checked against the actual structure.

The validity of available documents, particularly test results, needs to be assessed carefully and, where possible, checked against site observations.

A list of the required equipment and materials for the field inspection should be made, which may include the usual materials and equipment that are used for routine field inspections, such as

- hand lens,
- binoculars,
- crack gauge,
- powerful torch,
- measuring tape, and
- camera.

During the site inspection, visible signs of deterioration, including crack widths, directions, spacing and location should be recorded photographically as well as on copies of scale drawings or, at the very least, by means of freehand sketches with measured dimensions and directions, e.g. a north-point.

2.1.2 Observations on the structure

The site inspection of the affected structure should be designed for the type of structure concerned. A site investigation of a dam, for example, may be different from that of a bridge or a building structure, because of differing usage and structural aspects.

The environmental conditions to which the structure is exposed should be clearly described. In the case of a viaduct or a bridge in cold temperate climatic regions, clear distinctions should be made between the areas that are more exposed to de-icing salts and those that are not. Attention should also be given to situations where roof leakages, ponded water, groundwater, seawater or sea spray, de-icing salts, or condensation can occur.

2.1.3 Preliminary assessment of the site inspection

As AAR is a relatively uncommon phenomenon, in most cases, the site inspection will not result in AAR being suspected as a cause of the cracking. However, in some cases there will be evidence that suggests attack by AAR as the cause. It is likely that this will be the case in areas that have previously experienced problems with AAR.

The main external evidence of deterioration due to AAR is usually cracking typical of AAR (see Plates 1.1 to 1.7). In relatively unrestrained concrete, such as walls or structural pavements, the cracks usually occur in a characteristically random distribution or network, often referred to as 'map-cracking'. Plates 1.1 and 1.2 show examples of map cracking, although the cracks shown did not occur in unrestrained concrete. The expansive forces due to AAR may be restrained, for example, by the reinforcement. In that case, the cracks are usually preferentially oriented in the direction of the main reinforcement. Plate 5.9 shows an example of a longitudinal crack in a prestressed concrete cantilever, while Plate 2.1 shows the upper part of the structure shown in Plate 1.1, with the position and direction of the main tension over the Y support, clearly outlined by horizontal cracking. In fact, all of the cracking in the beam tends to be sub-horizontal. In contrast, Plate 2.2 shows map cracking in a massive support pier for a motorway structure. Here, the completely random cracking shows that the vertical stress in the column has not affected the direction of the cracks.

Other external features suggestive of damage due to AAR, such as surface discolouration, in particular along cracks, (see Plate 1.7), efflorescence and exudations, pop-outs and areas of structure with high moisture concentrations should be noted, recorded and photographed. Pop-outs, caused by expansion of individual near surface aggregate particles or of AAR-gel close to an exposed concrete surface, can sometimes cause detachment or "popping out" of a conical portion of the surface, leaving a small crater. Any exudations should be sampled by scraping off into a plastic bag which is then sealed, for later analysis.

2.1.4 Sampling of concrete

If the preliminary assessment strongly suggests that the observed cracking was caused by AAR, the next step is to take samples of the concrete for further examination and possibly for measurement of the concrete's physical properties. Sampling is usually by means of coring but in certain cases, e.g. with severe cracking, it may be possible to detach and bring lumps of concrete, defined by cracks, into the laboratory for testing.

Cores should be drilled with a diamond crown from locations that represent the various states of deterioration of a concrete structure including apparently sound concrete that seems not to have suffered AAR attack. During extraction of the cores, special attention should be given to the state of any reinforcement intersected, particularly the presence of corrosion and the condition of the bond of the steel to the concrete. Powdered concrete drillings may be taken for chemical analysis.

The number of samples that will be required depends on the type of structure and its complexity. The sampling scheme should provide a series of samples representative of the elements of the structure under investigation, as well as the various degrees of deterioration.

Cores and samples must be carefully labelled, showing the location in the structure from which they were taken, as well as giving notes as to the visual appearance of the concrete at their locality, as in the following example:

Structure sampled: Slip-and-slide interchange (31°28′40″S, 28°14′12″E), westbound carriageway, N1 highway

Date cored: 11 September 2001

Sample no. & length	Location	Description
IS – 200 mm	left lane, 2 m from E abutment I m north of kerb	apparently sound concrete
ID – 165 mm	as for IS, but next to kerb	core taken over longitudinal crack, recovered in 4 pieces labelled I to 4 from surface down
2S – 200 mm	right lane, from W abutment I m north of kerb	apparently sound concrete
2D – 120 mm	as for 2S, but next to kerb	taken over intersection of 2 cracks, recovered in 6 pieces, labelled I to 6 from surface down

2.2   PETROLOGY OF AAR-SUSCEPTIBLE MINERAL AND ROCK TYPES

Much of the material given in this section is based on Alexander and Mindess (2005) with supplementary material from Larbi, et al. (2004).

It was noted in Section 1.6 that three conditions are necessary for AAR to develop and progress:

- a source of alkalis – usually the cement or binder itself,
- the presence of reactive aggregates, and
- an environment that supplies sufficient moisture to the concrete to trigger and sustain the AAR reaction.

All three conditions are simultaneously necessary for AAR to develop. Alkalis and their types have been described in Section 1.6, as well as the environments conducive for AAR to occur. This section is mainly concerned with the types of reactive aggregates that are involved in AAR, and diagnostic tools to identify them. The petrology of AAR-susceptible rock types will be described first. This will be followed by diagnostic tests and their interpretation and some practical guidance on their use in specific applications.

Petrology is the study of rocks, while petrography refers to the methods and techniques of examination, imaging and analysis of rocks and rock-forming minerals. Both are important in considering AAR in concrete. AAR occurs when certain alkali-reactive components of aggregates are present in a concrete mix under enabling environmental conditions. The **minerals** in the rock type from which the aggregate is derived are of concern. These minerals can exist in different crystalline or microstructural states so that the presence of a mineral, determined by chemical composition alone, is not an indication of possible AAR-susceptibility or of AAR itself. The minerals must be thermodynamically reactive, that is they must be able to dissolve and break down in the presence of strong alkalis and subsequently form an expansive gel. A key element in any investigation of the likely susceptibility of an aggregate source to AAR is therefore the identification of the mineral types and their microstructure.

2.2.1   Mineral constituents

Rocks may contain many different types of minerals, defined as 'naturally occurring inorganic substances of more or less definite chemical composition and usually of a specific crystalline structure' (ASTM C294, Alexander and Mindess, 2005). Rocks are classified in terms of their various origins, which in general terms may be volcanic, sedimentary, or metamorphic. Rocks are earth materials, and this implies that there will be a large variability in their properties, both spatially and temporally in any given deposit or quarry. Thus, aggregate producers need to be constantly vigilant as to the AAR-susceptibility of their rock types, particularly if certain of the rock types are known to be potentially reactive, and the rock or gravel is variable.

The nature of the primary mineral constituents commonly found in concrete aggregates is summarised in Table 2.1. They vary greatly from highly stable quartz to sometimes rapidly decomposing ferromagnesians. (The chemical nature of these materials is discussed in detail by Alexander and Mindess, 2005).

Table 2.1 Minerals found in concrete aggregates (based on information in ASTM C294).

Mineral	Examples
Silica	**Quartz** A very common hard mineral composed of silica (SiO_2). Pure quartz is colourless and glassy, without visible cleavage. Highly resistant to weathering. Abundant in sands, gravels, many sandstones, and many light coloured igneous and metamorphic rocks. Some strained or microcrystalline quartz may be alkali-reactive.
	Opal A hydrous form of silica, generally without crystal structure. It is usually highly alkali-reactive.
	Chalcedony A fibrous form of quartz with submicroscopic porosity. Frequently occurs with chert and is usually alkali-reactive.
	Tridymite and Cristobalite High temperature crystalline forms of silica associated with volcanic rocks, and alkali-reactive.
Feldspars	These alumino-silicate minerals are the most abundant rock forming minerals in the earth's crust, and are important constituents of most major rock groups. Feldspar minerals are differentiated by chemical composition and crystallographic properties.
	Potassium Feldspars: orthoclase and microcline
	Sodium Feldspars: albite, plagioclase} Intermediate feldspars are oligoclase, andesine,
	Calcium Feldspars: anorthite} labradorite, and bytownite.
	Potassium and sodium feldspars occur typically in igneous rocks such as granites and rhyolites; calcium feldspars occur in igneous rocks of lower silica content such as andesite, basalt and gabbro.
Ferromagnesian minerals	These are constituents of many rocks, comprising dark minerals, generally silicates of iron or magnesium or both.
	Amphiboles: e.g. hornblende
	Pyroxenes: e.g. augite
	Olivines: e.g. forsterite. Found only in dark igneous rocks without quartz.
	Dark micas: e.g. biotite, phlogopite. Easily cleave into thin flake and plates. See also below.
Micaceous minerals	These have perfect cleavage in one plane, splitting into thin flakes. Micas are common in all rock types, and as minor or trace constituents in many sands and gravels.
	Muscovites: Colourless to light green
	Biotites: Dark coloured to black
	Lepidolites: Light coloured to white
	Chlorites: Dark green coloured
	Vermiculites: Formed by alteration of other micas, brown coloured
Clay minerals	These are layered silicate minerals, the particle size range being less than 1 μm, e.g. hydrous aluminium, magnesium and iron silicates with variable cations such as calcium, magnesium, potassium, sodium, etc. Formed by alteration of other silicates and volcanic glass. They are major constituents of clays and shales, also found in altered and weathered igneous and metamorphic rocks, as seams and lenses in carbonate rocks, and as matrix or cementing material in sandstones and other sedimentary rocks. Rocks containing large amounts of clay minerals are generally soft and unsuitable for use as aggregates. Different types of clay minerals are frequently interlayered.

(Continued)

Table 2.1 (Continued).

Mineral	Examples
	Kaolinites, illites, and chlorites: These are relatively stable clay minerals, but are absorptive.
	Smectites and montmorillonites: These comprise the swelling clays, and are highly unstable volumetrically. If included in concrete they give rise to high volume changes on wetting and drying.
Zeolites	Zeolites are a large group of hydrated alkali-aluminium silicates, soft and light coloured. Usually formed from hydrothermal alteration of feldspars. Can contribute releasable alkalis to concrete through cation exchange. Some varieties give substantial volume change with wetting and drying. These minerals are therefore not favoured in concrete aggregates. They are rare except in basalt cavities.
Carbonate minerals	**Calcite:** Calcium carbonate, $CaCO_3$ **Dolomite:** Calcium and magnesium carbonate, $CaCO_3 \cdot MgCO_3$ Both are relatively soft minerals, soluble in acid.
Sulphate minerals	**Gypsum:** Hydrous calcium sulphate, $CaSO_4 \cdot 2H_2O$, or anhydrite, $CaSO_4$. Typically forms a whitish coating on sand and gravel, and is slightly soluble in water. Other sulphates, e.g. sodium and magnesium, can also be present. All sulphates can attack concrete and mortar.
Iron sulphide minerals	**Pyrite, marcasite, and pyrrhotite:** Frequently found in natural aggregates. They may oxidize to sulphuric acid, and form iron oxides and hydroxides, which can attack or stain the concrete.
Iron oxides	**Magnetite:** Black coloured common mineral, Fe_3O_4 **Haematite:** Red coloured common mineral, Fe_2O_3 **Ilmenite:** Black, weakly magnetic, less common mineral, $FeTiO_3$ **Limonite:** Is a brown weathering product of iron-bearing minerals These minerals give colour to rocks and also colour concrete. They are frequently found as accessory minerals in igneous rocks and sediments. Magnetite, ilmenite, and haematite ores are used as heavy aggregates.

Note: Minerals are defined as naturally occurring inorganic substances of more or less definite chemical composition and usually of a specific crystalline structure (ASTM C294 definition).

While Table 2.1 gives information on minerals in concrete aggregates, Table 2.2 summarises the major rock and mineral types that are alkali-reactive, indicating the reactive component in each case. The table indicates that silica minerals (which include volcanic glasses) and dolomite are the only minerals with a widely established record of being alkali-reactive (Dolar-Mantuani, 1983). Despite the fact that alkali reactive minerals are limited, these minerals are essential constituents of a large number of rock types. Also, all three major rock classifications contain alkali-reactive rocks, and therefore the problem of AAR can occur in virtually any region of the world – a fact noted in Section 1.1 and attested by documented occurrences of the phenomenon in, for examples: Argentina, Australia, Brazil, Canada, China, Denmark, England, France, Iceland, Ireland, Mozambique, Netherlands, New Zealand, South Africa, Switzerland and U.S.A. However, concrete can be used successfully as a construction material even when AAR does occur (despite precautions), as will be shown in Chapters 4 and 5.

Table 2.2 Minerals, rocks and other substances that are potentially deleteriously reactive with alkalis in concrete (Oberholster, 2001, Alexander and Mindess, 2005).

Minerals
Opal

Tridymite

Cristobalite

Chalcedony, cryptocrystalline, microcrystalline or glassy quartz

Coarse-grained quartz that is intensely fractured, granulated and strained internally or rich in secondary inclusions

Siliceous, intermediate and basic volcanic glasses

Vein quartz

Rocks

Rock		Reactive component
Igneous	Granodiorite Charnockite Granite	Strained quartz; microcrystalline quartz
	Pumice Rhyolite Andesite Dacite Latite Perlite Obsidian Volcanic tuff	Silicic to intermediate silica-rich volcanic glass; devitrified glass, tridymite
	Basalt	Chalcedony; cristobalite; palagonite; basic volcanic glass
Metamorphic	Gneiss Schist	Strained quartz; microcrystalline quartz
	Quartzite	Strained and microcrystalline quartz; chert
	Hornfels Phyllite Argillite	Strained quartz; microcrystalline to cryptocrystalline quartz
Sedimentary	Sandstone	Strained and microcrystalline quartz; chert; opal
	Greywacke	Strained and microcrystalline to cryptocrystalline quartz
	Siltstone Shale	Strained and microcrystalline to cryptocrystalline quartz; opal
	Tillite	Strained and microcrystalline to cryptocrystalline quartz
	Chert Flint	Cryptocrystalline quartz; chalcedony; opal

(Continued)

Table 2.2 (Continued).

Rocks

Rock	Reactive component
Diatomite	Opal, cryptocrystalline quartz
Argillaceous dolomitic limestone and calcitic dolostone	Dolomite; clay minerals exposed by dedolomitisation
Quartz-bearing argillaceous calcitic dolostone	

Other Substances
Synthetic glass;
silica gel

Notes
1　Reactive aggregates vary widely in their reactivity depending on geological origin, location within a given geological formation, and location within a given source such as a quarry. Thus, where an aggregate may be suspected of being alkali-reactive, it is necessary to test the specific source from which it is derived.
2　Only dense reactive aggregates are potentially damaging; porous rocks, even if reactive, generally contain sufficient pore volume to absorb the expansive gel.
3　Rocks listed above although being siliceous in character may be innocuous if their siliceous minerals are not alkali-reactive.
4　Alkali-susceptible rocks are found worldwide, and there are few countries or regions that have not had AAR problems in concrete. Specific detail on suspect rocks and aggregate sources must be sought in the region or country concerned. (e.g. CSA A23.1–00 – Appendix B, Pike, 1990), Oberholster, 2001, Hobbs, 1987, Swamy, 1997.

2.2.2　The alkali-silica reaction

The alkali-silica reaction, or ASR is governed by the nature and quantity of reactive silica present. Table 2.2 indicates rocks and minerals that are susceptible to alkali attack. In ASR, **reactivity** of the silica is important. Glassy, amorphous silica such as opal and chalcedony are highly reactive, while crystalline varieties such as stable crystalline quartz are not. Reactive silica can occur as poorly crystallised minerals (cryptocrystalline or microcrystalline) or as strained quartz crystals, which appear in rocks subjected to shearing and distortion from tectonic forces often resulting in metamorphism, or where intrusive igneous rocks have induced re-crystallisation of existing sedimentary or metamorphic rocks. It is fairly easy to identify strained silica (or strained quartz) lattices using polarised light in a petrographic microscope. (See Plate 2.2.)

Thus, reactivity of silica increases with decreasing crystallinity. There is a rough correlation in laboratory tests between the amount of microcrystalline quartz present and mortar bar expansion (see Section 2.3.2). Microcrystalline quartz consists of randomly orientated grains with a mean grain size $<100 \times 10^{-6}$ m. Microcrystalline quartz may be formed either by crystallisation from amorphous material, e.g. opal, or by recrystallization from highly strained and deformed quartz due to stress metamorphism. The latter results in crystal lattice dislocations which enhance reactivity (Grattan-Bellew, 1992). However, reactivity also depends on the texture and origins of the quartz grains. The small grain size implies that solubility of the quartz is enhanced, thereby favouring ASR.

Deleterious expansion resulting from in ASR depends also on the **amount** of reactive silica, in conjunction with the nature of the reactive aggregate. For example, highly reactive forms of silica such as opal may require only 2 per cent by mass to cause deleterious expansion, while less reactive varieties such as greywackes require in excess of 20 per cent to be problematic (Oberholster, 2001, Alexander and Mindess, 2005).

A phenomenon termed the "pessimum effect" occurs with certain rapidly reacting forms of silica, notably opal. This refers to a critical aggregate content, as little as 1 per cent in some cases, at which measured expansion is greatest. The pessimum proportion is related to the reactivity of an aggregate, with more reactive aggregates showing lower pessimum proportions. For certain reactive cherts, the pessimum proportion may vary from 5 per cent to as high as 50 per cent by mass of an aggregate. The pessimum effect can also refer to a critical alkali content for a given aggregate content at which expansion is a maximum. Reasons for the pessimum effect relate to the quantity of reactive components being either too small or alternatively sufficiently large to dilute and reduce the harmful effects of potential expansion. For many other conventional and more slowly reacting aggregate types such as reactive quartzites and greywackes, a pessimum effect is not observed.

The question of where the reactive silica occurs – whether in fine or coarse aggregate – is also important. Experience suggests that the larger size fractions play the dominant role. This may be a case of the pessimum effect whereby the finer fractions do not contribute significantly to internal damage by virtue of their much higher surface area which distributes the expansive sites and dilutes the reactive effect. This is similar to the effect generated by highly reactive forms of finely divided silica (e.g. silica fume or fly ash) which reduce or prevent ASR by inducing multiple reaction sites that effectively immobilise the alkalis.

2.3   ASSESSING AGGREGATES FOR AAR-POTENTIAL

In most cases of AAR testing, engineers are concerned with assessing the susceptibility of aggregates to AAR. When considering a diagnostic test programme for AAR, the objectives of testing need to be clearly defined. This will also govern the nature of the investigation that is undertaken in terms of the tests likely to be used. Certain tests may give useful information on the likelihood of ASR occurring (e.g. ASTM C1260) but standard tests usually provide no useful information for the engineer in terms of how to avoid AAR or how to interpret the likelihood of continuing AAR damage to a structure.

The majority of tests are aimed at assessing the potential of aggregates for AAR. However, these tests may not be particularly useful in predicting the rate and degree of damage that is likely to occur in a real structure made with such aggregates. Further, many tests are broad 'categorisation tests' which are intended to screen aggregate sources in terms of AAR susceptibility. Aggregates that 'fail' such tests could possibly still be used successfully in concrete structures provided the proper precautions are taken. Thus, any testing programme must be carefully constructed to provide the maximum amount of useful information to the engineer. Typically this will involve not only an assessment of the potential of an aggregate for AAR, but also an assessment of the conditions under which the aggregate might still be used. Further, the nature of the investigation in terms of the tests selected is critical. For example, tests selected for screening a new source of aggregate might be very different to those for carrying out investigative work on a deteriorating structure. For other critical applications, detailed information will be required such as the threshold alkali content below which the aggregate is unlikely to show deleterious expansion. The nature of the assessment must be linked to the desired information and to the criticality of the assessment for the structure in question. The issue is complicated by differing national standards and test methods in which criteria and approaches may differ.

Table 2.3 gives a summary of aggregate tests for AAR. The tests are considered in three main categories:

- initial non-quantitative screening tests (used to make a provisional assessment),
- indicator tests to differentiate between potentially reactive and innocuous aggregates, and
- performance tests, giving information on limiting alkali contents to avoid damaging expansions. These three divisions follow a natural progression in many cases of diagnosis for AAR: initial indications on the likelihood of susceptible aggregates being present; more detailed tests to assess whether the aggregate is indeed alkali-susceptible; and tests that provide information on likely performance of a given concrete mix or set of concrete materials for a specific application.

The three categories of tests indicated above are discussed further below.

2.3.1   Initial screening tests

These are essentially qualitative tests used to make a preliminary assessment of aggregate-susceptibility to AAR. The use of aggregate petrography is of chief concern. This is a very powerful tool in the hands of an experienced petrographer, and in view of its importance and usefulness, is discussed in some detail in Section 2.4. The use of petrography is essential to confirm whether cracking in a structure is AAR-related or not. It is also very useful for an initial indication of the likelihood of AAR occurring with a given aggregate source, by the process of identifying possibly alkali-reactive minerals in the aggregate. The cost of a petrographic examination is miniscule in comparison with that of having to face the consequences of AAR developing in a structure.

2.3.2   Indicator tests

The purpose of these tests is to differentiate between potentially reactive and innocuous aggregates. Various versions of accelerated mortar bar test (e.g. ASTM C1260) are very commonly used worldwide, having gained international acceptance. The method involves monitoring expansion of mortar bars or prisms containing the test aggregate and immersed in a 1M sodium hydroxide solution at 80°C. The monitoring

Table 2.3 Aggregate assessment for AAR (excluding carbonate rock assessment).

Test designation and purpose	Material tested	Procedure and duration	Assessment criteria and test outcomes	Test standards	Limitations and remarks
I INITIAL SCREENING TESTS (Non- or Semi-Quantitative)					
Petrographic Examination. *Petrographic examination procedures for aggregates, according to standard descriptive nomenclature, as an aid to determine their performance*	Coarse and fine aggregates, or rock cores	Standard petrographic techniques, including optical microscopy, XRD analysis, differential thermal analysis. Presence and quantities of deleterious components such as: **Minerals:** opal, tridymite, cristobalite, chalcedony, chert **Rocks:** crypto-and microcrystalline quartz; evidence of deformation of quartz, such as undulatory extinction, intergrowth, or reaction with matrix. Also evidence of silica gel formation, cracking of aggregates and matrix, reaction rims, formation of crystalline silicates.	Standard petrographic techniques, including optical microscopy, XRD analysis, differential thermal analysis. Presence and quantities of deleterious components such as: **Minerals:** opal, tridymite, cristobalite, chalcedony, chert **Rocks:** crypto-and microcrystalline quartz; evidence of deformation of quartz, such as undulatory extinction, intergrowth, or reaction with matrix. Also evidence of silica gel formation, cracking of aggregates and matrix, reaction rims, formation of crystalline silicates.	ASTM C294 & C295 BS 812: Part 104 BS 7943 RILEM TC191-ARP (AAR-1)	These tests are for identification of potentially reactive constituents, and characterisation of minerals making up concrete aggregates. They are essential to confirm whether cracking in a structure is AAR-related or not – see ASTM C856 (Note at end of this table).
Gel-Pat Test *Detection of reactive silica, e.g. opaline silica*	Gravel-sized particles	Particles embedded in a cement pat, at 20°C or at 80°C; examined for 10 days for signs of reaction	Evidence of AAR gel and reaction	Appendix to BS 7943	Does not require sophisticated equipment, and can be performed on site.
2 RAPID INDICATOR TESTS (to determine whether aggregates are potentially reactive or innocuous)					
Potential alkali reactivity of aggregates (chemical method). *Determination of potential reactivity of siliceous aggregates*	Siliceous aggregates, crushed and sieved (150–300 μm)	Aggregates placed in 1M NaOH solution at 80°C for 24 h. Analysed for dissolved silica and reduction in alkalinity	Results checked against calibration curve; pass/fail criterion (deleterious or potentially deleterious)	ASTM C289	Not recommended for aggregates such as greywacke, hornfels, quartzite, granite, etc. Test shows some aggregates to be innocuous when they are known to have a poor service record.

Potential alkali-reactivity of cement-aggregate combinations (mortar-bar method). *Determination of susceptibility of cement-aggregate combinations to expansive reactions with alkalis*	Cement-aggregate combinations. Particular size fractions <4.75 mm required, obtained if necessary by crushing	Storage of 25 × 25 × 285 mm mortar bars at 38°C and 90% R.H., 3 months to 1 year. Essential to control R.H. at sufficiently high level during test	Criteria given in ASTM C33: Harmful reactivity if expansion >0.05 per cent at 3 months or >0.10 per cent at 6 months. Revised criteria for quartz-bearing rocks: Deleterious if expansion >0.05 per cent in 52 weeks. See Brandt & Oberholster (1983)	ASTM C227	Results take 3 months to 1 year. Quartz-bearing rocks require the longer test period. Useful test to determine susceptibility of combinations of cements and aggregates to harmful expansion. Reactive dolomitic aggregates not revealed by this test.
Accelerated mortar bar (mortar prism) test.* RILEM TC 106: "Ultra-accelerated mortar-bar test") *Determination of potential for deleterious alkali-silica reaction of aggregate in mortar bars*	Mortar prism comprising susceptible aggregates, with specified grading from 4.75 mm to 150 μm, obtained by crushing as necessary	Prisms 25 × 25 × 285 mm stored in 1M NaOH at 80°C for 14 days	Expansion after 14 days (12 days in South Africa): <0.10 per cent – non-expansive 0.10–0.20 per cent – slowly reactive or potentially reactive ≥0.20 per cent – deleteriously reactive CSA: >0.15 per cent – potential deleterious expansion	ASTM C1260 BS DD 249 CSA: A23.2–25 A RILEM TC 191-ARP (AAR- 2) SANS 6245	Based on work at NBRI in South Africa, Oberholster & Davies (1986); also used in USA and Canada (Hooton & Rogers, 1992), and recommended by RILEM. BS draft version. Rapid test, useful for slowly reacting aggregates or those producing expansions late in the reaction; generally reliable and reproducible, but not reliable for aggregates containing more than 2 per cent porous flint.

(Continued)

Table 2.3 (Continued).

3 PERFORMANCE TESTS (e.g. to provide information on limiting alkalis or structural performance)

Test designation and purpose	Material tested	Procedure and duration	Assessment criteria and test outcomes	Test standards	Limitations and remarks
Concrete Prism Method *Determination of the potential AAR expansion of cement-aggregate combinations*	Concrete aggregates proposed for actual construction. Na_2O_{eq} content in test = 5.25 kg/m³	Prisms 75 × 75 × 300 mm stored over water (100% R.H.) at 38°C. Cement and alkali contents stipulated in ASTM C1293 and other standards. Duration 3 months to 1 year	Any combination of cement and aggregate giving expansion after 52 weeks: <0.05 per cent – non-expansive (CSA: <0.04 per cent) 0.05–0.10 per cent – potentially or moderately expansive (CSA: 0.04–0.12 per cent) >0.10 per cent – expansive (CSA: >0.12 per cent)	ASTM C1293 BS 812: Part 123 CSA A23.2–14 A RILEM TC191-ARP (AAR-3)	Advantage that actual mixes can be tested in proportions specified. Can be used to evaluate effect of supplementary cementitious materials, and to assess alkali carbonate reactive aggregates. Long test duration required for meaningful results.
Modified Concrete Prism Method *Determining the potential AAR expansion of cement-aggregate combinations*	A Canadian test; procedures are identical to the conventional Concrete Prism Method immediately above, except that storage is at 60°C and 100% R.H. for a period of 3 months (91 days).			CSA A23.2–14 A (ASTM C1293)	Alternative rapid test to CSA A 23.2–14 A. Good correlation with standard test for sedimentary and carbonate rocks.
Ultra-accelerated concrete prism test	Concrete prisms containing aggregates proposed for construction	Prisms 75 × 75 × 250 mm stored over water at 60°C. Duration at least 20 weeks; longer for some slowly-reacting aggregates	Expansion after 3 months: <0.02 per cent – non-expansive <0.02 per cent after 6 months indicates minimal risk of AAR Shape of expansion curve to be considered also.	RILEM TC191-ARP (AAR-4)	Used to assess reactivity performance of particular concrete mixes

Long-term structural monitoring	Actual structures and structural members	Procedures designed for particular structures; generally involve monitoring expansions, deflections, and cracking with time; full-scale load testing may also be carried out. Criteria depend on particular structure. Excessive expansions, deflections or cracking taken as appropriate criteria.	Inst. of Struc. Engineers (1989); RILEM TC 20-TBS (1984)	On occasions, structural monitoring is the only way to assess the performance of an AAR-affected structure and to assess its on-going integrity

Notes:

- Aggregates differ in their alkali-reactivity. Furthermore, the demarcation lines between reactive and innocuous aggregates are not sharp; therefore judgement must be exercised when evaluating an aggregate.
- Other tests are: ASTM C342: Potential volume change of cement-aggregate combinations – *determines the potential AAR expansion of cement aggregate combinations (primarily used for aggregates from Oklahoma, Kansas, Nebraska, and Iowa)*; ASTM C441: Effectiveness of mineral admixtures or GGBS in preventing excessive expansion of concrete due to AAR – *determines effectiveness of supplementary cementing materials in controlling AAR expansion. (Also covered in CSA Standard A23.2–28 A)*; ASTM C856: Petrographic examination of hardened concrete – *outlines petrographic examination procedures for hardened concrete, useful in determining condition or performance.*
- ASTM approved a further test in 2004: C1567, Test Method for Determining the Potential Alkali-Silica Reactivity of Combinations of Cementitious Materials and Aggregate, Accelerated Mortar Bar Method. This test evaluates pozzolans and slag for controlling alkali-silica reaction, and can be used to determine the level of extender required to control AAR with a particular aggregate. It is a modification of test method C1260, which is strictly an aggregate test. Test results are produced in 14 days.
- RILEM tests have been produced by Technical Committees (TC) of RILEM. The TC previously concerned with AAR was RILEM TC 106 (2000), and is currently (2004) RILEM TC 191-ARP.
- RILEM TC 106 (2000) recommends that chemical testing such as ASTM C289 only be used as a secondary method, due to difficulties of interpretation.
- RILEM TC 191-ARP envisages that the AAR-4 test (Ultra-accelerated concrete prism test) may be used in three modes: for testing potential reactivity of an aggregate combination; as an ultra-accelerated version of the AAR-3 test, i.e. as a performance test for assessing the alkali-reactivity of a particular mix; and as a test for establishing the critical alkali threshold of a particular aggregate combination.
- Developments in BS standards on AAR depend on parallel developments in European Standards. For example, RILEM AAR-2 may in due course be included as a European Standard, replacing BS DD 249, for which no steps are presently being taken to convert it into a formal British Standard.
- Canadian Standards Association document CSA A23.1–00, Appendix B, contains useful information on AAR testing specific to Canadian conditions.
- The above table is based on an outline given by Oberholster (2001).

References consulted:
ASTM, CSA, and RILEM sources quoted in table; Oberholster (2001); Jones & Tarleton (1958); Brandt and Oberholster (1983); Oberholster and Davies (1986); RILEM TC106-AAR (2000); Sims and Nixon (2003a, b&C).
*Most commonly used test world wide.

period is usually 12 to 14 days. The method derives from work carried out in South Africa by Oberholster and co-workers (Oberholster, 1983).

ASTM C1260 is the basis of the test accepted and developed by RILEM TC 106 (AAR-2) and included in various national test methods. In addition to being the standard test used in the US, it appears in similar or identical form in the British, Canadian and South Africa standards. It is useful as a rapid indicator but does not necessarily remove the need for conventional concrete prism testing (e.g. BS 812: Part 123), which is covered below. In 2004, ASTM approved a modified C1260 test, C1567, which can be used to determine the level of pozzolan or slag required to control alkali-silica reactivity with a particular aggregate.

Other tests in this category need to be carefully judged as to whether the results are applicable in any given situation, since in some cases aggregates with a known poor service record may show up in the tests as innocuous. Some tests can take a very long time (up to 1 year) and are therefore not particularly useful. All of the tests require a high degree of skill to perform. Tests that take a long period of time are obviously more expensive.

A further test, based on a Chinese procedure and following similar protocols as ASTM C1260, is the 'Concrete Microbar Test' (Grattan-Bellew et al., 2003). It uses $40 \times 40 \times 160$ mm bars, with aggregate graded between 12.5 mm and 4.75 mm, water/cement ratio of 0.33, and immersion in a 1M NaOH solution for 30 days at 80°C. The method is claimed to be applicable to both alkali-carbonate and alkali-silica reactive aggregates (differentiated in the test by replacing a portion of the portland cement by a supplementary cementing material). Moderate correlations were found between 30-day expansions in this test and 1-year expansions measured in the modified concrete prism test.

2.3.3   Performance tests

These tests are used to assess the reactivity performance of particular concrete mixes. Thus, greater reliance can be based on the results in terms of a specific project. A further advantage is that they can be used to evaluate the effect of supplementary cementitious materials on potential AAR. A drawback is that in general at least three months is required in order to obtain meaningful results. Use of the test therefore requires careful pre-planning.

Particular mention should be made of long-term structural monitoring of AAR-affected structures. These represent the ultimate in performance tests and are dealt with in detail in Chapter 4.

2.3.4   RILEM technical committee contributions

RILEM (Reunion Internationale des Laboratoires et Experts des Materiaux), through its various technical committees, has made significant contributions to the understanding of and testing for AAR. RILEM Technical Committee TC 191-ARP, the successor committee to Committee TC 106, has developed a scheme for the integrated assessment of AAR, illustrated in Figure 2.2 (Nixon and Sims, 1996, Sims and Nixon, 2001, Sims and Nixon, 2003). Various elements are proposed:- petrographical examination (AAR-1); rapid testing using the accelerated mortar bar expansion test (AAR-2); and

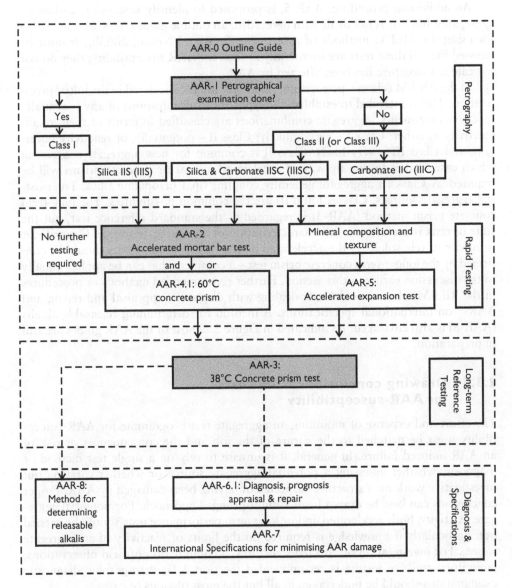

Figure 2.2 RILEM TC 106/TC 191 Integrated AAR Assessment Scheme (adapted from Sims and Nixon, 2003a, b). Shaded boxes represent those sections that had been published (2010).

a concrete prism expansion test (AAR-3) which is viewed as a long-term reference test. These three elements can readily be identified with the scheme in Table 2.3. An ultra-accelerated performance test for concrete is also envisaged (AAR-4) in which concrete prisms are stored at 60°C. It is intended that this test might be appropriate as a project-specific performance test, in that the actual concrete mix for the job is tested. International trials as a recommended method indicate that it is able to distinguish between non-reactive and reactive combinations, and is reproducible.

An additional procedure, AAR-5, is proposed to identify susceptible carbonate aggregates, but this is not dealt with here. Also, an outline guide (AAR-0) is available for using the RILEM methods of assessment (Sims and Nixon, 2003b). It must be stressed that all these tests are useful only as tests for AAR susceptibility; they do not indicate if a structure has been affected by AAR.

In the RILEM scheme, petrographic examination is considered in the initial investigation. This is intended to establish the types and concentrations of any potentially reactive constituents. Aggregate combinations are classified in terms of their alkali-reactivity as either: Class I – very unlikely; Class II – potentially, or reactivity uncertain; and Class III – very likely. Class II is common for new aggregate sources, in which case further testing by way of accelerated mortar or concrete prisms will be required. A Class III aggregate generally contains opal or opaline silica. For existing aggregate sources, Classes I and III are the common possibilities. The long-term concrete prism method (AAR-3) is regarded as the standard reference test, but the suite of tests is useful for a full characterisation of an aggregate source. Generally, it is unwise to rely solely on the accelerated mortar bar test and results should be confirmed by the longer-term concrete prism test – a comment that can be applied equally to the discussion earlier in this section. Further recommended methods or procedures in the RILEM scheme are AAR-6, dealing with diagnosis, appraisal and repair, and AAR-7 on international specifications. A method for determining releasable alkalis (AAR-8) is also envisaged. As indicated in Figure 2.2 most of the AAR guides are still in preparation.

2.3.5   Drawing conclusions from tests for AAR-susceptibility

The effort and expense of mounting an aggregate test programme for AAR-susceptibility must be matched to the nature of the job and the consequences or risk of an AAR-induced failure. In general, it is unwise to rely on a single test method or diagnosis, whether assessing a potential aggregate source or whether carrying out investigative work on a structure suspected of having been damaged by AAR. Proper conclusions can best be drawn from a multi-pronged approach. For potential aggregate reactivity, both accelerated and/or long term performance tests should be carried out, particularly if knowledge is required on the limits of reactivity of an aggregate source. For investigation of affected structures, a combination of field observations, laboratory and site testing is recommended (Section 2.1). Petrographic diagnostic examinations should be undertaken in all but the most obvious of cases.

2.4   AGGREGATE PETROGRAPHY

This section provides a background to aggregate petrography, and indicates how petrography can be used to characterize concrete aggregates and their properties.

Petrography is a powerful tool in the diagnosis and study of AAR in concrete. In the hands of a skilful petrographer, it can provide the information essential for a proper understanding of the cause of deterioration, and assist in indicating how the problem can be handled. Interpretation of petrographic results is always very important, and if the petrographer does not have a strong grounding in cement and concrete science, the engineer has to apply particular care in interpreting the results.

Petrography can assist in the following ways in regard to concrete aggregates: it can:

- describe aggregate constituents and determine their relative amounts, permitting classification,
- determine certain physical and chemical characteristics that may influence the performance of the aggregate in concrete,
- detect potentially deleterious constituents or undesirable minerals and contaminants in the aggregate, such as sulphates, as well as the presence of aggregate coatings,
- establish the likely performance of aggregates from new or untried sources by comparing their characteristics with aggregates of known performance, and
- assist in interpreting results of other standard aggregate tests and in selecting further tests for determining aggregate performance.

2.4.1   Petrographic composition and examination of aggregates

The petrographic composition of an aggregate describes its mineralogical constituents. Petrographic examination provides information on both the nature and properties of the constituent rocks and minerals, including the presence and amount of any undesirable constituents and the degree of weathering. Petrographic examination should be performed for new and tested sources of aggregates or for existing sources that are highly variable or where a different rock type is encountered. Petrographic examination also indicates the limitations of the aggregate for use in concrete. However, it can also sometimes permit an otherwise unsuitable material to be beneficiated to avoid or eliminate undesirable portions or constituents.

A correct description of aggregate composition includes both the overall composition of the rock and mineral types present, and the nature of the minerals in the individual grains. The more complex the rock type, the more information is required from the analysis. The degree of weathering of the aggregate minerals also needs to be identified, since weathering can give rise to secondary or tertiary minerals that may be unsuitable in concrete (for example active clay minerals such as smectites). From the perspective of AAR, it goes without saying that the examination should identify and quantify potentially alkali-reactive constituents and suggest further tests to determine their activity.

The microtexture and microstructure of the aggregate should be described since this may have a strong influence on aggregate performance. In the context of AAR, this would include aspects such as the presence of the straining of quartz crystals, micro- or crypto-crystallinity, undulatory extinction, etc.

A good petrographic analysis therefore provides an aggregate evaluation that identifies aspects of engineering significance and the possible effects of the various constituents on the properties of concrete, and not merely a petrological description.

Plate 2.3 gives an example of the use of petrography to detect an alkali-reactive aggregate, in this case a greywacke from South Africa. This rock is well known for its susceptibility to AAR. The problematic constituent is strained quartz (crystals, arrowed in the Plate 2.2) which is revealed by the use of crossed polars in a petrographic microscope.

2.4.2 Analysis techniques

Examination and analysis techniques involve optical microscopy of polished and thin sections and of grain mounts in immersion oils from which photomicrographs can be made. Techniques also include X-ray diffraction (XRD) analysis, differential thermal analysis (DTA), infrared spectroscopy, scanning electron microscopy (SEM) with energy-dispersive X-ray analysis (EDX), and back-scattered electron imaging (BSE) of polished sections.

Polished sections can provide useful information about AAR damage in cores or samples removed from structures, such as the nature of the aggregate and pattern of cracking. Impregnating the section with an epoxy resin containing a fluorescent dye can enhance visualisation of the crack pattern. When the surfaces are properly polished, they can be readily examined under low power optical microscopes or simply, with the aid of an UV-light system. Under UV illumination, the dye mixed with the epoxy resin fluoresces which permits identification of crack patterns and reaction products in the cracks. Other features that can easily be examined or identified on the polished section are:

*   crack patterns in aggregate and cement paste and fillings in cracks,
*   intensity of internal cracking of aggregate and cement paste, presence of AAR-gel in relation to cracking, and
*   quality of the cement paste: weak, fractured, poorly bonded to particles and any other irregularities in the concrete.

Standard petrographic guides are ASTM C295: "Standard Guide for Petrographic Examination of Aggregates for Concrete" and BS 812: Part 104:1994: "Testing Aggregates". These standards describe procedures for qualitative and quantitative petrographic examination of aggregates.

These two documents are valuable guides for the petrographer and equally for the concrete engineer who wishes to understand and interpret the results of petrographic analyses. The guides outline the extent to which petrographic techniques should be used, the selection of properties that should be looked for, and the manner in which such techniques should be employed in aggregate sample examination. They provide information on minimum sample sizes and size fractions and techniques such as particle counting etc. Guidance is given for examination of both coarse and fine aggregates, with thin-section examination invariably being required for fine aggregate. The documents also provide guidance on proper reporting of petrographic analyses (in this regard, see Smith and Collis, 2001 and Mielenz, 1994). The report should contain at least the following:

*   identification of the sample regarding its source and proposed use,
*   test procedures used,
*   data on composition and properties of the aggregates, e.g. particle shape, texture, and possible coatings,
*   description of the nature and features of each important constituent of the sample,
*   identification of possible unfavourable effects in concrete of any deleterious constituents, and
*   recommendations for any additional petrographic, chemical, physical, or geological investigations that may be required.

2.4.3   Assessing residual ultimate expansion of concrete in structures

Engineers who investigate AAR-affected structures often pose the question of what expansion potential still remains in the concrete of the structure. This question cannot be answered as the answer depends almost entirely on the climatic conditions in which the structure is situated, the conditions of restraint imposed by reinforcing and structural geometry and future changes in the use and degree of exposure of the structure. Certain authors, e.g. Swamy (1997) suggest accelerated laboratory expansion tests on cores treated in various ways, e.g. immersion in 4% NaCl solution at 38°C, and simultaneous field expansion tests on cores exposed in the same environment as the structure. The results of accelerated tests of this type are almost certainly meaningless. Even the tests on cores, released from the stress-field of the structure, and exposed to only a small sampling of the environment are unlikely to provide any meaningful information in a quantitative sense.

REFERENCES

Alexander, M & Mindess, S 2005, *Aggregates in concrete*, Taylor and Francis, London, U.K.

ASTM C294 (Current version), *Standard descriptive nomenclature for constituents of concrete aggregates*, ASTM, West Conshohocken, U.S.A.

ASTM C295 (Current version), *Standard guide for petrographic examination of aggregates*, ASTM, West Conshohocken, U.S.A.

ASTM C1260 (Current version), *Standard test method for detection of alkali-silica reactive aggregate by accelerated expansion of mortar bars*. ASTM, West Conshohocken, U.S.A.

BS 812: Part 123: 1999, *Method for determination of alkali-silica reactivity: Concrete prism method*, British Standards Institution, London, U.K.

BS 812: Part 104: 1994, *Testing aggregates*, British Standards Institution, London, U.K.

Brandt, MP & Oberholster, RE 1983, *Investigation of Tygerberg aggregate for potential alkali reactivity*, CSIR Research Report BRR 500 (in Afrikaans language), CSIR, Pretoria, South Africa.

CSA A23.1–00 (Current version), *Concrete materials and methods of concrete construction/methods of test for concrete*. Canadian Standards Association, Ottawa, Canada.

Dolar-Mantuani, L 1983, *Handbook of concrete aggregates*, Noyes, Park Ridge, NJ, U.S.A.

Grattan-Bellew, PE 1992, 'Microcrystalline quartz, undulatory extinction and the alkali-silica reaction', *9th Int. Conf. on AAR in Concrete*, London. U.K. pp. 383–394.

Gudmundsson, G & Asgeirsson, H 1975, 'Some investigations on alkali-aggregate reaction', *Cement and Concrete Research*, vol. 5, no. 3, pp. 211–220.

Guédon, JS & Le Roux, A 1994, 'Influence of microcracking on the onset and development of the alkali silica reaction', *3rd CANMET-ACI Int. Conf. on Durability of Concrete*, Nice, France, pp. 713–724.

Hammersley, GP 1992, 'Procedures for assessing the potential alkali-reactivity of aggregate sources', *9th Int. Conf. on AAR in Concrete*, London, U.K., pp. 411–417.

Hobbs, DW 1987, 'Some tests on fourteen years old concretes affected by the alkali-aggregate reaction', *7th Int. Conf. on AAR in Concrete*, Ottawa, Canada, pp. 342–346.

Hooton, RD & Rogers, CA 1992, 'Development of the NBRI rapid mortar bar test leading to its use in North America', *9th Int. Conf. on AAR in Concrete*, London, pp. 461–467.

Jones, FE & Tarleton, RD 1958, *Reactions between aggregates and cements*, Parts V and VI, Alkali-Aggregate Reaction, National Building Studies, no. 25, HMSO, London, U.K.

Larbi, J, Modry, S, Katayama, T, Blight, G & Ballim, Y 2004, 'Guide to diagnosis and appraisal of AAR damage in concrete structures: The RILEM TC 191-ARP approach', *12th Int. Conf. on AAR in Concrete*, Beijing, China, pp. 921–932.

Mielenz, RC 1994, 'Petrographic evaluation of concrete aggregates', in *Significance of tests and properties of concrete and concrete making materials*, ASTM STP 169C: pp. 341–364, American Society for Testing and Materials, Philadelphia, U.S.A.

Nixon, P & Sims, I 1996, 'Testing aggregates for alkali-reactivity', *Materials and Structures*, vol. 29, no. 190, pp. 323–334.

Oberholster, RE 1983, 'Alkali reactivity of siliceous rock aggregates: diagnosis of reactivity, using cement and aggregate and description of preventive measures', *6th Int. Conf. on AAR in Concrete*, Copenhagen, Denmark, pp. 419–433.

Oberholster, RE 2001, 'Alkali-silica reaction', in *Fulton's concrete technology*, eds. BJ Addis & G Owens, 8th edn, Cement and Concrete Institute, Midrand, South Africa.

Oberholster, RE & Davies, G 1986, 'An accelerated method for testing potential alkali reactivity of siliceous aggregates', *Cement and Concrete Research*, vol. 16, pp. 181–189.

Pike, DC 1990, *Standards for aggregates*, Ellis Horwood, Chichester, U.K.

Shepherd, R & Frost, JD (eds) 1995, *Failures in Civil Engineering*, American Society of Civil Engineers, Virginia, U.S.A.

Sims, I & Nixon, P 2001, 'Alkali-reactivity – a new international scheme for assessing aggregates', *Concrete*, vol. 35, no. 1, pp. 36–39.

Sims, I & Nixon, P 2003a, *Towards a global system for preventing alkali-reactivity. The continuing work of RILEM TC 191-ARP*, American Concrete Institute Special Publication SP-212, American Concrete Institute, Michigan, U.S.A., pp. 475–487.

Sims, I & Nixon, P 2003b, 'RILEM – recommended test method AAR-O: detection of alkali-reactivity potential in concrete – outline guide to the use of RILEM methods in assessments of aggregates for potential alkali-reactivity', *Materials and Structures*, vol. 36, pp. 472–479.

Sims, I & Nixon, P 2003c, 'RILEM – recommended Test Method AAR-1: detection of potential alkali-reactivity of aggregates – Petrographic Method', *Materials and Structures*, vol. 36, pp. 480–496.

Smith, MR & Collis, L 2001, *Aggregates: sand, gravel and crushed rock aggregates for construction purposes*, Engineering Geology Special Publications 17, Geological Society, London, U.K.

Stanton, TE 1940, 'The expansion of concrete through reaction between cement and aggregate', *Proc. American Society of Civil Engineers (ASCE)*, vol. 66, pp. 423–431.

Swamy, RN 1997, 'Assessment and rehabilitation of AAR-affected structures', *Cement and Concrete Composites*, vol. 19, pp. 427–440.

PLATES

Plate 2.1 Example of cracking concentrated along the line of main reinforcing and also showing sub-horizontal tendency of other cracks.

Plate 2.2 Completely random directions of AAR cracking that is not affected by major stress directions.

Plate 2.3 Micrograph showing presence of strained quartz crystals, (arrowed).

Chapter 3

Effects of AAR on engineering properties of concrete – results of laboratory determinations

3.1 LABORATORY SPECIMENS AND CORES TAKEN FROM STRUCTURES

The descriptions of the AAR phenomenon and process, in the first two chapters, have been of an observational nature that is essential for diagnosis and also for understanding the AAR process and its superficial effects. For engineering decision-making, a quantitative knowledge of the effects of AAR as it progresses, as well as the extent to which it can be expected to progress, is required. In other words, quantity must be added to quality. There are two basic ways of acquiring this knowledge, each with its own size scale and appropriate timing in the course of an investigation. These are laboratory and field or in situ tests. Laboratory tests are usually performed on small specimens taken from full-size structures, whereas in situ tests are carried out on the actual structure. In an intermediate form of test, large components can be taken from a structure and brought into a laboratory to be tested.

Laboratory tests have the advantages that they are relatively low in cost and can be multicated to allow of statistical analysis. They can also be applied to specific parts of an affected structure. Another major advantage is that if the tests involve stressing or straining as they usually do, stress, strain and boundary conditions can be better defined and controlled. If environmental and loading variables such as water content, temperature, load duration and time rate of load application are considered important, these too can be controlled in the laboratory. In relation to time effects, specimen size can be scaled (within acceptable limits) to allow slow time effects (e.g. leaching or permeability) to be measured within a manageable time scale.

Laboratory tests also have many disadvantages. Among these are:

- The largest size of specimen that can be managed in the laboratory may be too small for realistic measurements to be made, or engineering properties to be realistically assessed. For example, to reliably establish the properties of concrete, the minimum dimension of a specimen should be at least five times the size of the largest aggregate particle. Even for relatively small 19 mm aggregate, this requires making or taking and testing 100 mm diameter cylinders or cores.
- It may be difficult to obtain cores from the structure being investigated that are statistically representative of the in situ concrete. This may be due to difficulties of access, or difficulty in anchoring the coring drill to the structure, or because the concrete is so fissured by AAR cracks that intact cores cannot be recovered.

In the latter case, an intact core will usually not be representative of the mass of concrete in the structure. In any case, recovered, testable cores will be biased towards representing the better concrete composing the structure, not the average concrete, and certainly not the worst.

- The concrete may prove to be so variable in properties that it becomes prohibitively expensive or too time-consuming to recover enough cores and to do enough tests to give a realistic and statistically reliable assessment of the concrete's properties.

In all cases, it is best to adopt a staged approach, in which an initial limited, exploratory program of coring and laboratory tests is undertaken. This will usually expose any of the three difficulties described above, and possible others too. If the results of the limited program are re-assuring, it may be decided that the information gained, together with an ongoing, carefully focused, supervised, analysed and regularly reported monitoring program will take care of any future safety concerns. Clearly, if the results of the preliminary test program are less than re-assuring, further tests will be required, possibly including in situ, in-service measurements on the structure, or full-scale loading to the design load. The latter two courses of action will be described, with examples, in Chapter 4.

3.2 THE PROCESS OF CRACKING

Figure 3.1a shows the basis for a simple analysis to assess the possibilities involved in initiating longitudinal or transverse surface cracks in an unreinforced concrete prism. The prism measures B × H × L externally. The heart of the concrete, that is initially unaffected by drying of the surface, measures b × h × l. This will be referred to as the heartcrete. The outer layer of concrete, the covercrete, usually loses water by evaporation and tends to shrink relative to the heartcrete which restrains the shrinkage. This will set up compressive stresses in the heartcrete and opposing tensile stresses in the covercrete. Assuming that the compressive stress σ_c and the tensile stress σ_t are

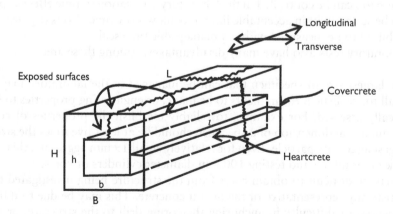

Figure 3.1a Basis for investigating likely surface cracking modes.

uniform throughout the thicknesses of the covercrete and heartcrete, the opposing Compressive (C) and Tensile (T) forces will be given by:

Longitudinally: Transversely:
$C_L = bh\sigma_{cL}$ $C_T = bL\sigma_{cT}$ or $C_T = hL\sigma_{cT}$

The tensile forces in the covercrete will be given by

Longitudinally: Transversely:
$T_L = (BH - bh)\sigma_{tL}$ $T_T = (B - b)L\sigma_{tT}$ or $T_T = (H - h)L\sigma_{tT}$
                                       There are thus two possibilities for T_T as these
                                       two values need not be the same.

For C and T to balance:
$bh\sigma_{cL} = (BH - bh)\sigma_{tL}$ $bL\sigma_{cT} = (B - b)L\sigma_{tT}$ or $hL\sigma_{cT} = (H - h)L\sigma_{tT}$

If the prism is square, B = H and b = h, and
Longitudinally: Transversely:
$\sigma_{tL}/\sigma_{cL} = b^2/(B^2 - b^2)$ $\sigma_{tT}/\sigma_{cT} = b/(B - b) = h/(H - h)$
If b = 9/10B   If b = 9/10B
$\sigma_{tL}/\sigma_{cL} = 4.26$ $\sigma_{tT}/\sigma_{cT} = 9$
If b = 4/5B   If b = 4/5B
$\sigma_{tL}/\sigma_{cL} = 1.78$ $\sigma_{tT}/\sigma_{cT} = 4$

In both cases, it is more likely for cracking to occur where the ratio σ_t/σ_c is greater, i.e. longitudinal cracking at right angles to the transverse σ_t is more likely than transverse cracking. Therefore a square prism is more likely to crack longitudinally than transversely.

For a slab-like prism:
$\sigma_{tL}/\sigma_{cL} = bh/(BH - bh)$ $\sigma_{tT}/\sigma_{cT} = b/(B - b)$ or $\sigma_{tT}/\sigma_{cT} = h/(H - h)$

If the ratio of H to B, is 1/10, if B = 3000 mm then H = 300 mm and if (B − b) = 2900 mm, then (H − h) = 100 mm, taking the same covercrete thickness of 50 mm all round.

$\sigma_{tL}/\sigma_{cL} = 2900 \times 200/(3000$ $\sigma_{tT}/\sigma_{cT} = 2900/(3000 - 2900) = 29$
$\times 300 - 2900 \times 200)$

or

$\sigma_{tL}/\sigma_{cL} = 1.81$ $\sigma_{tT}/\sigma_{cT} = 200/100 = 2.0$

Of the three possibilities, σ_{tT}/σ_{cT} for b/(B − b) = 29 would give the most likely source of cracking. Hence the most likely crack would be horizontal, and parallel to the face of the slab. In the case of a reinforced slab the tendency to crack parallel to the face would be increased by the imposition on the concrete of bond stresses in the plane of the reinforcing. The combined effect of shrinkage and bond stresses could result in delamination of the covercrete from the heartcrete.

Hence by this simple analysis, a longitudinal crack will always (theoretically) be more likely to occur. However, in reality the initiation and control of crack direction is affected by many other factors, for example, the direction, proximity to the surface and diameter of reinforcing, bond stress and also temperature effects relating to variations in exposure to sun, etc. The prediction of the analysis is, however, borne out by delamination of cover layers and horizontal cracking observed in, e.g. concrete road pavements. (See Section 4.10.4 and Plates 3.2 and 4.14.)

3.3 DIFFERENCES BETWEEN LABORATORY SPECIMENS AND CORES TAKEN FROM AAR-AFFECTED STRUCTURES

As far as possible, when discussing the effects of AAR on concrete in terms of laboratory tests, the results of tests on cores taken from AAR-affected structures will be used to maximize the reality of the discussion and conclusions. In discussing certain properties, however, it will be necessary to refer to the properties and behaviour of laboratory-made concrete specimens when relevant tests on specimens taken from AAR-affected structures are not available.

One of the main reasons for adopting this approach is that, in reality, AAR-damaged concrete forming part of a reinforced or mass concrete structure, is only ever sampled and tested after signs of AAR damage appear. This may happen 10 to 15 years after completion of construction. The intake towers at the Itezhitezhi dam in Zambia illustrate one exception to this time scale (Thaulow, 1983). The dam was completed in 1976 and by early 1980 a substantial swelling of the concrete in the intake towers, of the order of 40 mm in a height of 30 m, was noticed. Over the next two years a further 10 mm of swelling was measured to give an estimated 50 mm over about 5 years or an average of 0.3×10^{-3} per year. After a year of observation, measurements showed that the expansion had ceased, but it is not known whether or not it ever resumed. Most laboratory research-type tests use means to accelerate the initiation and progress of the AAR so that the AAR-affected concrete can be tested months or at most a year or two after casting. In these specimens, whether cylinders, prisms, model beams or columns, the AAR process occurs by consumption of the water contained by the concrete at casting. Depending on the method used to accelerate the AAR, available water is augmented by absorption from the specimens' surrounding water bath or humid atmosphere. In contrast, a structural member is cast and cured in its formwork for a period of days. After stripping and depropping, the concrete surface is exposed to the atmosphere and usually dries, causing surface shrinkage. As described in Section 3.2, the outer layer of concrete, or covercrete, therefore develops a tensile stress that puts the inner, or heartcrete, into compression. In the long term, the AAR proceeds by consuming the pore water the concrete already contains, augmented by externally sourced water such as seepage from faulty drainage ducts, impinging rain, etc. Eventually, even if no shrinkage cracking occurs, the AAR causes the heartcrete to expand or swell until the covercrete cracks, resulting in the familiar open tension cracks of an AAR-damaged concrete surface.

The differences in the expansive strains involved in laboratory specimens and full-scale structures can be very large. The first surface cracks can be expected to appear when the tensile strain at the surface exceeds 100 to 150×10^{-6}. Thereafter the cracks will widen, often seasonally, and new cracks will appear when the local surface tensile strain exceeds the cracking limit, as shown by the crack width measurements in Figure 3.1b. These crack measurements were made in Johannesburg, where the June to October period is dry, hence crack opening stops during this period and resumes when the rains come in October or November. Occasionally, a crack may stop widening (e.g. crack 7 in Figure 3.1b) and new cracks (e.g. 10 and 11) will appear and open.

It is usually impossible to assess the strain corresponding to a given crack opening, because the length of concrete in which the expansive strain is represented by the crack width is not known. The British Institution of Structural Engineers' report (1992) suggests that the crack widths along at least five straight lines at least 1 m long and at least 250 mm apart be summed and divided by the total length of the lines to give an estimate of the swelling strain in a particular direction. In principle, this method could be tested out on small linear components such as concrete balustrades or railway sleepers, but it is doubtful if it could be used or tested meaningfully on a large structure.

However, overall strains in laboratory specimens and in cores taken from structures and tested in the laboratory can be directly compared. As examples, Figure 3.2 compares the growth with time of swelling strains measured in restrained and partly restrained swelling of laboratory concrete prisms stored over a water surface (Berra, et al., 2008), with similar strains measured in cores drilled from a 27 year old structure and stored over water in exactly the same way (Alexander, et al., 1992). The only differences for the two sets of specimens were that the lateral dimensions of the laboratory concrete prisms were 75×75 mm while the cylindrical cores were 80 mm in diameter. Also, the temperature for the prisms was maintained at 38°C while that for the cores was kept at 23°C. The free expansions of the two specimens differ by almost a factor of 10. Differences in water permeability between two specimens may allow

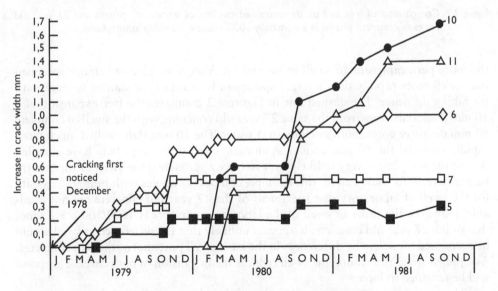

Figure 3.1b Growth of surface crack widths with time. Observations taken on column supporting motorway structure, from time when cracking was first noticed.

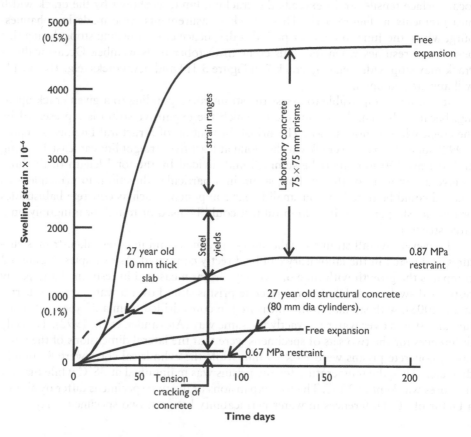

Figure 3.2 Comparison of free and partly restrained swelling of laboratory prisms and 27 year old cores of concrete placed in a nominally 100% relative humidity atmosphere.

the more permeable one to swell more rapidly. Also, a smaller or thinner specimen may swell more rapidly than a larger specimen, because of the shorter seepage path for taking up water. The dashed line in Figure 3.2 compares the free expansion of a 10 mm thick slab sawn from the same 27 year old concrete, with the swell of complete 80 mm diameter cores, both swelling over water. The 10 mm slab swelled much more rapidly than did the 80 mm core. The ultimate swell of the core may have equalled that of the slab, but is very unlikely ever to have approached the swell of the laboratory concrete. To some extent, this example compares "apples with oranges" as data for the swell of laboratory cast specimens of the 27 year old concrete are not available. However, the order of swell of the laboratory concrete is 4600/500 = 9.2 times that of the 27 year old concrete. It appears unlikely that the difference in swell would be accounted for solely by differences in the basic swell potential of the two concretes. An applied stress restraining the swell reduces the expansion in both cases. This point will be returned to later.

The various sections in the remainder of this chapter will describe properties of AAR-affected concrete that are the product of physical tests on specimens taken from concrete structures and tested in the laboratory. Most of the properties are derived from measurements of the strain response to loads applied to the concrete either with or without eventual destruction. The laboratory tests inevitably require the use of measuring instruments. As the use of these is often common to both laboratory and field tests, but field tests often require techniques and instruments not used in the laboratory, measuring instruments, their principles and their application will be described in Chapter 4, Section 4.8. The reader of Chapter 3 is referred to Section 4.8 for details of the instruments mentioned in the present chapter.

3.4   THE TESTING OF CORES AND LABORATORY-PREPARED CYLINDERS OR PRISMS

This section will describe the more common tests and set out some basic rules that should be followed when testing cores drilled from an AAR-affected structure or specimens prepared in the laboratory. The limitation of overall specimen size in relation to the maximum size of aggregate has already been mentioned in Section 3.1. To yield realistic measurements of strength or elastic properties, the minimum dimension of the specimen should not be less than five times the maximum size of aggregate. The 5:1 ratio is an absolute minimum, and where possible, the ratio should exceed this value.

3.4.1   Stresses in a cylinder subject to compression between rigid platens

Figure 3.3a shows the calculated stresses in a cylinder of elastic material loaded axially with stress Q between perfectly rough, rigid end plates (Filon, 1902). The roughness of the end plates resists the radial Poisson strain of the cylinder, causing restraining stresses to be generated. In the figure, $\bar{\sigma}_A$ is the average actual axial stress in the cylinder, which is close to Q between values of z/H between 0 and about ± 0.8. $\bar{\sigma}_R$ and $\bar{\sigma}_T$ are the average radial and tangential stresses which would be zero in the absence of friction on the ends of the cylinder. It is clear from Figure 3.3a that only between z/H = 0 and ± 0.5 are the stresses close to those assumed in the usual uniaxial compression test, i.e. $\bar{\sigma}_A = Q, \bar{\sigma}_T = \bar{\sigma}_R = 0$.

If the friction on the loading platens is reduced by, e.g. bedding the cylinder on a thin sheet of rubber, then because the Poisson's ratio of rubber is larger than that of concrete, radial and tangential tensile stresses will be induced that will reduce the compressive failure stress. A similar effect will be caused if a material like soft board or a capping of a material less stiff than the concrete, like plaster of paris, is used.

End restraint on a cylinder subjected to compression also varies radially across the cylinder. For example, at the ends, σ_A at the perimeter of the cylinder increases to 1.65Q, σ_R reaches 0.9Q on the axis and 0.45Q on the perimeter and σ_T also becomes 0.9Q on the axis, although it is zero on the perimeter. At the centre of length of the cylinder (z/H = 0), σ_A varies from 1.14Q on the axis to 0.9Q on the perimeter, and σ_R and σ_T are zero. Hence even if H = 2D = 4a the restraint at the ends of the cylinder has a slight effect over the central half-length. Even though H = 2D is generally regarded as an acceptable proportion for a test cylinder, H = 2.5D is even better, and the British

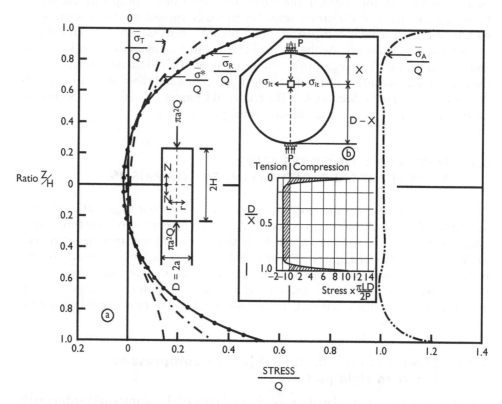

Figure 3.3 (a) Theoretical stresses in a cylinder of elastic material loaded axially with stress Q between perfectly rough, rigid end plates (Filon, 1902). (b) (inset on a) stresses in indirect or splitting tension test.

Institution of Structural Engineers (1992) recommends using a length of at least 2.5D. H should not be increased much beyond 3D, though, because of the possible introduction of buckling effects if the cylinder becomes too slender.

It is therefore best to test a concrete cylinder between smooth steel platens, make sure that the height to diameter ratio is at least 2.5 and make any strain measurements over the central half-length of the concrete specimen. Great care should be taken to prepare the ends of a core for compression testing by smooth sawing them at right angles to the core axis and, if necessary, grinding them smoother or planer.

All compression testing machines that are well designed have each of the two loading platens mounted on a spherical bearing designed to correct any small eccentricities of loading introduced by non-parallel ends of a specimen. However, these devices can only effectively correct small eccentricities, hence the need for careful proportioning of specimens for testing. Loading platens should always be leveled in two directions at right angles, using a small spirit level when setting up each specimen for testing.

3.4.2   Load-controlled and strain-controlled testing

Most machines for testing concrete are of the load-controlled type with the load applied by a hydraulic cylinder. As the load increases and the specimen compresses, the loading platen follows the compression so as to maintain the desired level and rate of loading. When the failure load is reached and the load reduces, the strain energy stored in the testing machine is released into the specimen. The result is usually a sudden catastrophic failure in which control over the load is lost and the post-failure stress-strain relationship cannot be measured.

This problem can be overcome by using a stiff, strain-controlled testing machine. The frame and mechanism of the machine is made very much stiffer than any specimen it is designed to test. This minimises the rate at which strain energy stored in the machine can be released as the specimen fails and the load decreases. The strain-control allows the specimen to be strained at a constant pre-set rate while the resisting force generated by the specimen is measured and recorded. This enables both the rising and falling (or pre- and post-failure) parts or branches of the complete stress-strain curve to be measured and recorded.

Strain-controlled machines are particularly suitable for measuring the stress-strain curve for concrete in tension, as it is possible to record the falling branch of the stress-strain curve.

3.4.3   Measuring the elastic modulus
and Poisson's ratio for concrete in compression

Figure 3.4a illustrates the experimental arrangement for measuring the elastic modulus E and Poisson's ratio ν in a laboratory test. The prepared test cylinder is instrumented with at least one pair of long gauge length (50 mm) electric resistance strain gauges (see Section 4.8 for details of these and other instruments) mounted vertically and diametrically opposite. These measure the axial compressive strain in a compression test. The reason for using a pair of gauges is to detect and compensate not only for the effects of possible eccentricity of loading, but also to compensate for possible differences in concrete properties in the specimen from one side to the other. Two pairs of longitudinal gauges are better than one, but one pair is a minimum requirement. An alternative arrangement is to measure the compression of the cylinder over its central half-height by means of dial gauges or LVDTs mounted in a cage supported in place by two pairs of diametrically opposite Demec targets glued to the cylinder a gauge length (usually equal to D) apart. (See Figure 3.4a).

The gauges can be connected in pairs as a half-bridge to give an average axial strain, or four gauges can be connected as a full bridge. (See Section 4.8 for details.) In the half bridge, a pair of diametrically opposite ERS gauges could be connected in the places marked "measuring gauge" and "dummy gauge" in Figure 4.6c to form a "half bridge". Alternatively, if two pairs of gauges are used, the second pair could be connected in place of the standard resistances R to form a "full bridge".

If measurements of Poisson's ratio are required, two long gauge length ERS gauges can be mounted wrapped around the circumference of the specimen, as shown in Figure 3.4a. These can be connected as a half bridge to give average values of the

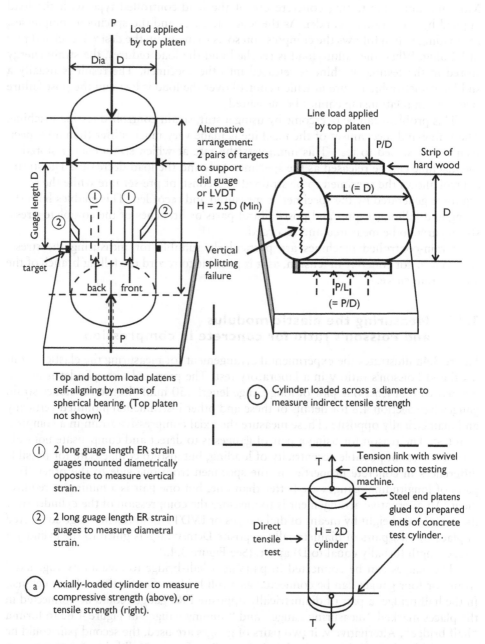

Figure 3.4 (a) Principle of direct compressive or tensile strength testing. (End plates glued to concrete cylinder and pulled to measure tensile strength). (b) Principle of indirect or splitting tensile strength test.

circumferential strain that corresponds to the axial strain. As the circumference is equal to πD, the circumferential strain equals the lateral strain.

Figure 3.5 shows a set of stress-strain curves for compression tests on cylinders of concrete that had been subjected to accelerated AAR expansion in laboratory tests (Wen, et al., 2000). These curves show the effect of increasing AAR expansion on the modulus (slope of the stress-strain curve) of the concrete. It also shows the Poisson ratio strains corresponding to the compressive strains. As the original work did not include measurements of lateral strain, the curves in Figure 3.5 have been calculated as −0.2 × the compressive strains to show the lateral strains schematically. The diagram also shows two common ways of expressing the modulus E. This can either be as a tangent modulus at the origin, which indicates the stiffness of the concrete at small strains, or as a chord modulus covering a particular stress range. In the illustration, the tangent modulus at the origin is 8.2 GPa, while the chord modulus over the stress range of 0 to 30 MPa is 7.3 GPa. Both are very low values, and indicate the extent to which AAR in these accelerated laboratory tests has caused the concrete to deteriorate.

3.4.4   Measuring the direct tensile strength

The direct tensile test is very similar to the compressive test (see Figure 3.4a). The prepared ends of the cylinder are glued to steel loading platens using a high-strength epoxy adhesive. Once the adhesive has cured (usually after 24 h at about 20°C) the specimen is set in the tensile testing machine, hanging from a top tensile link set in the upper jaws of the testing machine and with the bottom link in the lower jaws of the testing machine. Both links should have two way swivel connectors so that there is no eccentricity of the tensile loading. Failure strains are extremely small and it is difficult to measure E in tension, except by means of a stiff, strain-controlled machine. The failure does not always occur within the central length D of the specimen and

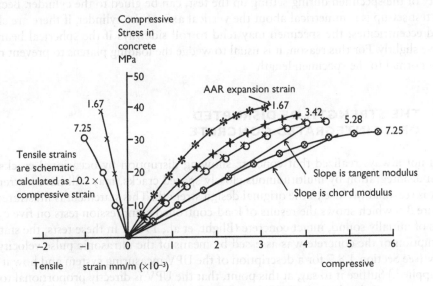

Figure 3.5 Stress – strain curves for concrete in compression (after Wen, et al., 2000).

results are usually more erratic than those of compression tests. (See, e.g. Figure 3.8c.) Unless the specimen fails by parting through one of the glue lines, a test is considered successful if the failure surface passes entirely through concrete, although the surface may be stepped or otherwise irregular.

3.4.5   Measuring the indirect or splitting tensile strength

Elastic analyses of a cylinder loaded by compressive line loads across a diameter (as shown in Figures 3.3b and 3.4b) show that most of the diametral line connecting the lines of action of the two line loads is subjected to an almost uniform tension. If the total line load is P and the length of the cylinder of diameter D is L (usually L is made equal to D), the line load per unit length (P/L) is related to the tensile stress across a diameter in line with the loads by:

$$\sigma_{it} = 2P/\pi DL \tag{3.1}$$

If $L = D$, $\pi DL = \pi D^2$ and since $\pi D^2/4 = A$, the cross-sectional area of the cylinder,

$$\sigma_{it} = P/2A \tag{3.1a}$$

(But, note that this applies only when $L = D$).

This test is also known as the Brazilian test as it originated in Brazil. It is very quick and simple to perform and the specimens need the minimum of preparation. The usual length to diameter ratio L/D is 1.0, but when testing cores, especially of AAR-affected concrete, it may be necessary or expedient to test cylinders of a length either less or more than D.

The loading strips can be of plywood or hardboard, and if found necessary for stability of the specimen during setting up the test, can be glued to the cylinder. Because the test set-up is symmetrical about the vertical axis of the cylinder, if there are slight load eccentricities, the specimen may tend to roll sideways if the spherical bearings move slightly. For this reason, it is usual to wedge the loading platens to prevent rotation normal to the specimen length.

3.5   THE STRENGTH OF DISRUPTED OR DISINTEGRATED CONCRETE

It is not always realized that concrete showing disruption by occasional cracks, or even disintegration by multitudinous closely spaced cracks, may still have a strength that is reasonably close to the original design strength. This statement is illustrated by Figure 3.6 which shows the results of load-controlled compression tests on five cylinders of initially sound, intact concrete (Blight, et al., 1981). In these tests, the state of disruption of the concrete was assessed by means of the ultrasonic pulse velocity, or UPV (see Section 4.8.7 for a description of the UPV measuring system and how it can be applied.) Suffice it to say, at this point, that the UPV is directly proportional to the square root of the dynamic modulus of elasticity of the concrete. Any disruption in

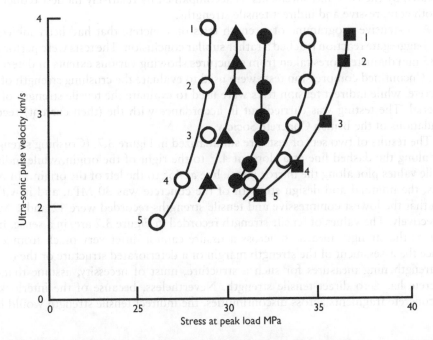

Figure 3.6 Effect of repeated load-controlled compression of initially sound, intact concrete cylinders to maximum (failure) load on ultrasonic pulse velocity (UPV).

the form of a crack running across the path of the UPV pulse will reduce the modulus, and this will also reduce the pulse velocity. The UPV of sound, strong concrete is usually 4 to 4.5 km/s.

The ultrasonic pulse velocity (UPV) was measured across the diameter of each concrete cylinder at mid-height. The cylinder was then loaded in compression to just past the peak load and was then unloaded and the ultrasonic pulse velocity re-determined. After this, the specimen was re-loaded to the maximum load it could sustain and then unloaded once more. The process was repeated four times, giving the five sets of five measurements shown in Figure 3.6.

The results show in each case, that as the concrete became progressively disrupted by the loading process, the ultrasonic pulse velocity decreased progressively to 50% or less than its initial value. The stress that the partly-disrupted concrete could sustain in each loading cycle did not however decrease by very much. The "best" and "worst" cases recorded in Figure 3.6 were:

"best": Initial UPV and strength: 3.95 km/s   29 MPa
     UPV strength on 5th loading: 1.90 (48%)   26 (90%)
"worst": Initial UPV and strength   3.90   37
     After 5th loading   2.05 (52%)   32 (86%)

Similar results were obtained in load-controlled indirect tension (Brazilian) tests where the UPV was measured axially. These tests show that severe disruption, as indicated by the UPV measurements, is accompanied by relatively modest reductions in both compressive and indirect tensile strengths.

An extensive programme of strength tests on concrete that had been subject to alkali-aggregate reaction reached a rather similar conclusion. The tests were performed on 85 mm diameter cores taken from structures showing various extents of deterioration. Unconfined compression tests were used to evaluate the crushing strength of the concrete, while indirect tension tests were used to evaluate the tensile strength of the material. The testing was carried out in accordance with the (then current) recommendations of the British Concrete Society.

The results of two sets of tests are summarized in Figure 3.7. (Crushing strengths plot along the dashed line at a slope of 45° to the right of the origin, while indirect tensile values plot along the mirror image line at 45° to the left of the origin.) In both cases, the nominal and design strength of the concrete was 30 MPa, and it will be seen that the lowest compressive and tensile strengths recorded were 12 and 1 MPa, respectively. The values of tensile strength recorded in Figure 3.7 are, in a sense, fictitious as the strength measured across a fissure cannot differ very much from zero. Hence the assessment of the strength margin of a deteriorated structure or the design of strengthening measures for such a structure, must of necessity, assume that the concrete has zero direct tensile strength. Nevertheless, because of the interlock of the concrete fragments across discontinuities, the indirect tensile strength would be a

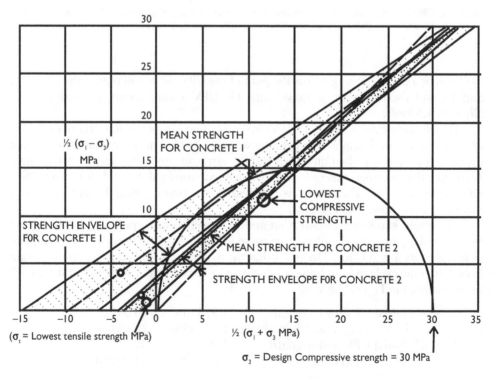

Figure 3.7 Strength envelopes established for two deteriorated concretes by means of compression and indirect tension tests.

valid measurement for assessing the shear strength across any surface subject to even a small compressive stress.

3.6 ELASTIC PROPERTIES, COMPRESSIVE, INDIRECT AND DIRECT TENSILE STRENGTHS OF AAR-AFFECTED CONCRETE

The elastic modulus E of concrete and Poisson's ratio v do not appear to have been given much attention in research into the properties of concrete from AAR-affected structures. However, Hasparyk, et al. (2004), as part of a survey of the properties of AAR-affected hydro-power plants in Brazil have made an interesting collection of data that are illustrated among other data in Figure 3.8. The rather scattered measurements were made on cores taken from different AAR-affected parts of four different power plants. (Paulo Afonso, AF 1, 2, 3 & 4) in Brazil. Ono (1989) assembled an even larger set of data on the elastic moduli and strengths of concrete, taken from a number of AAR-affected structures. In Ono's case the data came from bridge and retaining structures in Japan. Ono's data have been superimposed on the Hasparyk, et al., data in Figure 3.8a and the two sets of data largely coincide. Data for the Churchill dam (Oberholster, 1989, Section 5.8.2) have also been plotted on Figure 3.8a and also fit the general trend. Figure 3.8a shows that the ratio E/σ_c (where σ_c = compressive strength) can vary from 300 to 1000, with most of the data lying between 300 and 800, while Figure 3.8b shows that the ratio of σ_c/σ_{it} (σ_{it} = indirect tensile strength) can vary from 20 to 11.

A series of tests on cores taken from the AAR-damaged Kamburu spillway in Kenya (Sims and Evans, 1988), which compare compressive and indirect tensile strengths found that the ratio of σ_c/σ_{it} varied from 10 to 20. These results are also shown in Figure 3.8b.

Siemes and Visser (2000) made a study of the relationship between compressive strength of AAR-affected concrete and both indirect (σ_{it}) and direct σ_t tensile strength. Their data for indirect tensile strength are superimposed on Figure 3.8b and fall within the 20 to 11 limits. Their data on direct tensile strength σ_t fall well below the range for σ_{it}, and although all of their data was for concrete with σ_c of 50 MPa or more, the limits lie between σ_c/σ_t = 60 and 20.

Referring to Figure 3.7, it will be seen that a concrete having a uniaxial compressive strength of 30 MPa and a uniaxial (direct) tensile strength of 1 MPa (σ_c/σ_t = 30) could quite well fit the strength envelopes shown in the diagram.

The data in Figure 3.8 cover a number of structures, including strengths of unaffected concrete and strengths with various extents of AAR degradation. In a more closely focused study of a single AAR-damaged reinforced concrete pier and cap, Blight and Alexander (1988) found (see Figures 3.9a and b) that the ratio E/σ_c varied from 600 to 400 while the ratio σ_c/v (v = Poisson's ratio) varied from 130 to 260. The range of Poisson's ratio was from 0.19 to 0.29 and did not appear to depend on the strength of the concrete. In the case of the ratio E/σ_c, both highest and lowest ratios occurred with the same value of σ_c. The limits found by Blight and Alexander have been plotted on Figure 3.8a and agree reasonably well with those established by Ono (1989), Hasparyk, et al. (2004), Sims and Evans (1988) and Oberholster (1989).

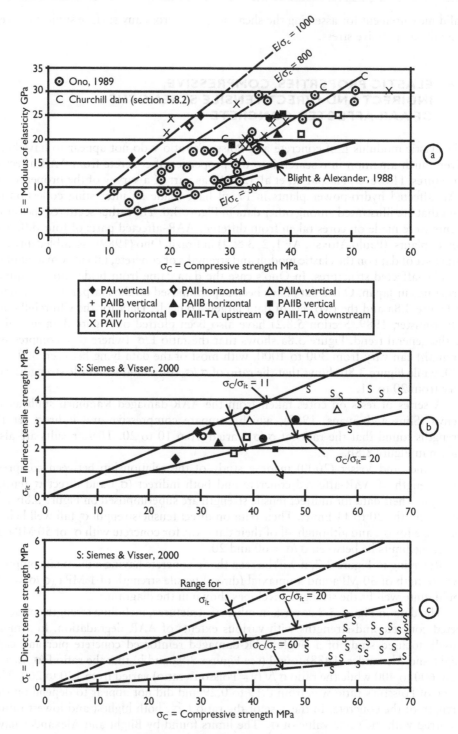

Figure 3.8 Relationships between compressive strength σ_c for AAR – affected concrete and modulus of elasticity E, indirect tensile σ_{it} strength and direct tensile strength σ_t.

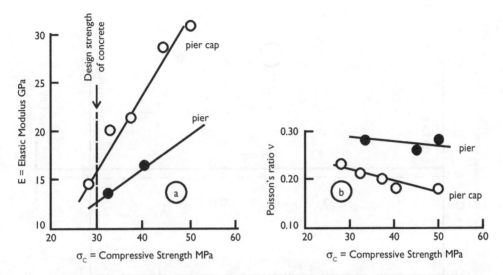

Figure 3.9 Relationship between elastic modulus, poisson's ratio and strength for AAR-affected concrete.

The ranges for the ratios E/σ_c, σ_c/v and σ_c/σ_{it} in Figures 3.8 and 3.9 are all wide, but provide a means of obtaining a first estimate of elastic (E and v) and indirect tensile strength (σ_{it}) properties from compressive strength tests, the perceived extent of deterioration of the concrete and the design strength, assuming that the design compressive strength equaled the original value for the now damaged concrete.

It is seldom that attempts have been made to assess the variation of the properties of an AAR-deteriorated structural member along its length or through its thickness. This was done in the case of pre-demolition tests carried out on the upper beam of a portal frame (Alexander, et al., 1992). The beam was due for demolition and rebuilding (see Section 5.6.3 and Figure 5.23). To assist in deciding the length of beam requiring demolition, three horizontal cores were taken along the vertical line indicated ("cores taken on this line") in Figure 5.23. The cores penetrated the full 1240 mm thickness of the beam and were drilled at vertical intervals of 400 mm in the 1800 mm depth of beam, within and missing the top and bottom longitudinal reinforcing and vertical stirrups by locating the reinforcement with a covermeter. The cores were divided into suitable lengths for testing, and after measuring the transverse UPV, were tested for elastic modulus E and crushing strength σ_c. The mean compressive strength was 31 MPa, compared with the design value of 30 MPa.

Figure 3.10a shows the UPV values plotted against the mean distance of each core from the north facing side of the beam. The lowest UPV values were 3.9 km/s and the maximum was 4.4 km/s. The mean value was 4.09 km/s, as compared with the mean value of 4.07 km/s for UPV's measured in situ three years earlier by direct transmission through the thickness of the beam. The UPV values were distributed randomly through the width (thickness) and depth of the beam, but the UPV limits of 3.9 to 4.4 km/s were both satisfactory values and, in themselves, did not indicate that the beam needed demolishing.

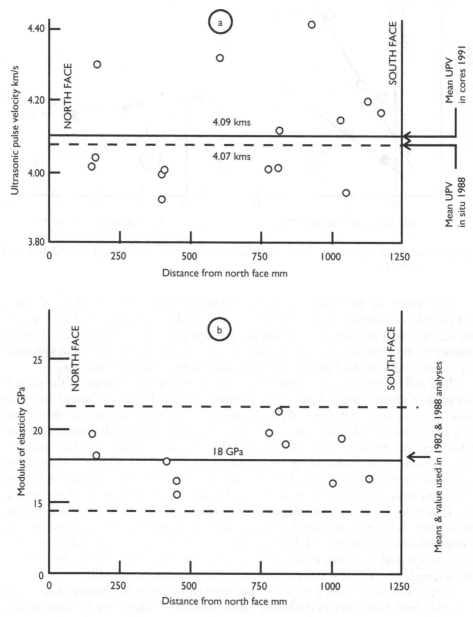

Figure 3.10 Distribution through the width and depth of a beam of (a) ultrasonic pulse velocity and (b) modulus of elasticity. Measurements were made on three cores drilled across the complete 1240 mm thickness of the beam of the portal frame shown in Figure 5.23.

Figure 3.10b shows the measurements of modulus of elasticity E. These all fell between the bounds of E/σ_c = 700 and 500, close to the limits quoted earlier (Blight and Alexander, 1988) of 600 to 400. As with the UPV values, the distribution of high and low values of E appeared to be random through the width and depth of the beam.

3.7 CREEP OF AAR-DAMAGED CONCRETE UNDER SUSTAINED LOAD

Remedial and strengthening measures for structures weakened by alkali-aggregate reaction, will often involve prestressing procedures. Hence it is necessary to have a knowledge both of the elastic and time-dependent properties of the material. Figure 3.11 summarizes the elastic and time-dependent deformation properties of six specimens of concrete. These were all cored from the same AAR-affected structure in 1980. Specimens A, B and C were tested in 1980 and the remaining three tests were performed in 1998 on the same group of cores. Specimen A was a concrete that by visual inspection was undamaged. Specimens B and C were damaged by AAR with C (visually) damaged somewhat more than B.

It will be noted that the elastic or instantaneous deformation of the deteriorated concrete, was approximately three times that of the apparently sound concrete, while the creep strain over a period of nine months (100 days) was two and a half to four

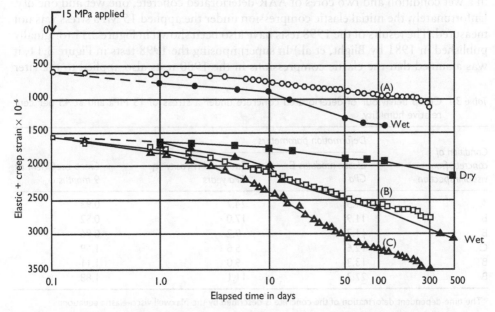

Figure 3.11 Typical creep curves under stress of 15 MPa at a relative humidity for 45% for:
1980 Tests A: Undeteriorated concrete ○   1998 Tests – Undeteriorated concrete: WET   ●
B: Deteriorated concrete □   Deteriorated concrete: DRY   ■
C: Deteriorated concrete △   WET   ▲

times as large. In making this comparison, however, it should be remembered that all three specimens were subjected to the same stress of 15 MPa. Whereas 15 MPa represented approximately 50 per cent of the strength of specimen A, it may have represented 80 or 90 per cent of the strength of specimens B and C. Also, these were laboratory tests on cores that had been completely stress-relieved by the coring process. Tests on in situ concrete may not show the same instantaneous strain. Indeed, in the only case of creep measured in the field (on record), the field creep was far exceeded by the creep measured on cores in the laboratory (Section 5.4.1, Figure 5.17.).

Table 3.1 summarizes the results of a number of Young's modulus and creep tests carried out in 1980 on concrete with different visual degrees of damage. Before these tests, the cores were dried out to a relative humidity of 45% and were maintained at this humidity during testing. It will be seen that deteriorated concrete, especially that in category B is characterized by a low modulus of elasticity, a low structural viscosity, as well as an unusually high creep coefficient (see Table 3.1). With regard to the creep coefficient (defined as the ratio of creep strain to elastic strain at a particular time) normal undamaged concretes would be expected to have values of the order of 0.5 which is less than most of the values listed in Table 3.1. The important point, however, is that both the elastic and time-dependent deformation properties of the concrete must be known when designing remedial measures for structures that have deteriorated by AAR, especially those involving prestressing.

After the tests recorded in Table 3.1, the cores were kept stored in a dry laboratory cupboard until 1998, when it was decided to retest them. The identification of the individual cores had been lost, but it was decided to test one core of sound concrete in a wet condition and two cores of AAR-deteriorated concrete, one wet and one dry. Unfortunately, the initial elastic compression under the applied 15 MPa stress was not measured. The results of the 1998 tests have also been shown in Figure 3.11 (originally published in 1981 by Blight, et al.) In superimposing the 1998 tests in Figure 3.11, it was assumed that the elastic compressions in the 1980 tests also applied to the later

Table 3.1 Creep behaviour of deteriorated concrete under a stress of 15 MPa and at 45 per cent relative humidity.

Condition of concrete from visual inspection	Deformation parameters		
	Elastic modulus E GPa	Structural viscosity* η GPa years	Creep coefficient at 9 months φ_9
A	31.5	24.7	0.93
B	11.9	17.0	0.52
B	11.7	9.2	0.96
C	12.6	5.6	1.69
B	13.3	9.0	1.11
B	27.3	11.1	1.88

*The time-dependent deformation of the concrete is described by the Maxwell visco-elastic equation:

$$\varepsilon(t) = \sigma \left(\frac{1}{E} + \frac{t}{\eta} \right) \tag{3.2}$$

in which $\varepsilon(t)$ = time-dependent strain, σ = sustained stress, t = time under stress, η = structural viscosity. A = visually undamaged; B = slight to moderate damage; C = moderate to severe damage.

tests. On this basis, there is reasonable agreement between the two sets of tests, with the wet tests undergoing slightly larger creep strains than the tests on dry cores.

3.8 THE EFFECTS ON EXPANSION OF COMPRESSIVE STRESS

Restraint of the expansion of concrete resulting from AAR may have several causes. It may arise from thrusts developed between abutting structural elements that are either both affected by AAR, or only one of the two may be affected. A good example is described in Section 5.8.3 in which Gocevski and Pietruszcak (2000) describe the structural distress caused by the restrained expansion of a mass-concrete water intake structure bearing against a mass concrete gravity dam. A common example is that of lengths of concrete parapet wall or road kerbs in which the expansion joints provided have closed as a result of AAR expansion and which buckle or go out of line under the resulting compressive stress.

The most common source of restraint, however, is that provided by reinforcing.

3.8.1 Restraint on expansion imposed by reinforcing

The British Institution of Structural Engineers (1992) have published a very comprehensive collection of data on the restraining effect of reinforcing which appears as Figure 3.12.

It will be seen from Figure 3.12a that expansion can be reduced by 70–90%, while stresses induced in concrete by the restrained reinforcing as high as 14 MPa have been measured. If one takes a steel/concrete modular ratio of 10, this would mean a maximum additional stress in the steel of 140 MPa. The results shown in Figure 3.12 are from laboratory tests and 14 MPa corresponded to an extremely high rate of expansion of >0.5 mm/day. Considering the much slower rates of expansion experienced in the field and the effects of creep and relaxation in the concrete, the opinion of the Institution of Structural Engineers was that the maximum induced compressive stress in concrete is likely to be about 4 MPa, or an induced steel stress of 40 MPa.

More recent work (Jones and Clark, 1996 and Smaoui, et al., 2004) found values of induced concrete stress that agreed with the lesser values shown in the original version of Figure 3.12 which adds weight to this opinion. The upper curve by Hobbs in Figure 3.12b should probably be disregarded, especially as it relates to an exceptionally highly accelerated rate of expansion.

Figure 3.12 shows that expansion is reduced as the steel percentage increases, but the induced stress in the concrete increases. On the other hand, if the reinforcement percentage is small and the expansion relatively large, the strain, and therefore the stress in the steel will be large. For example, if 0.5% of reinforcing is present, Figure 3.12a shows that the steel strain could be as high as 70% of the unrestrained expansion. If the free expansion strain is 0.1% (1×10^{-3}), (see Figure 3.2) the stress induced in the steel could be $70\% \times 0.1\% \times 200 \times 10^3 = 140$ MPa which gets back, if the modular ratio is 10, to the figure mentioned earlier of 14 MPa as a stress that could possibly be induced in a heavily reinforced concrete (4% steel and a rapid rate of expansion).

The work reported by Jones and Clark (1996) as well as Smaoui, et al. (2004) gives very high steel stresses at low percentages of reinforcing, as shown in Figures 3.12b

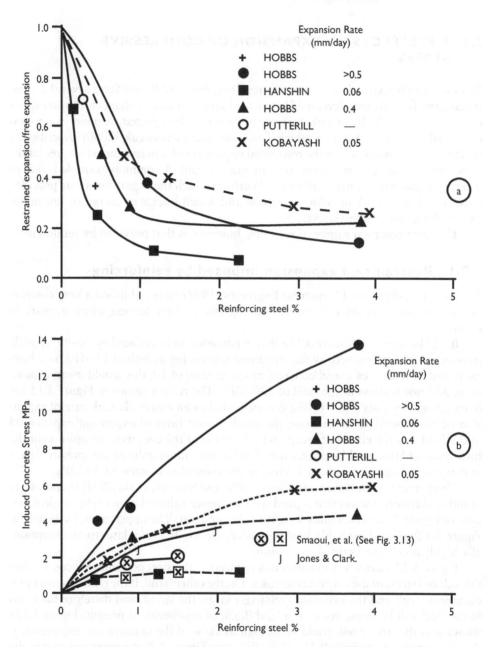

Figure 3.12 Effect of restraint by reinforcing on (a) expansion by AAR, (b) induced compressive stress in concrete structure, from time when cracking was first noticed.

and 3.13. Mild steel yields at about 250 MPa, hence any stress calculated from measured strain that equals or exceeds 250 MPa indicates that yield may have been reached in mild steel reinforcing. As Figure 3.13 shows very clearly, this can possibly happen when reinforcing of less than 1% of the concrete area is used in AAR-affected concrete.

3.8.2   Restraint on expansion imposed by adjacent structures or structural elements

As mentioned at the start of this section, concrete subject to AAR may be restrained by adjacent concrete, against which it abuts and which either partially or completely

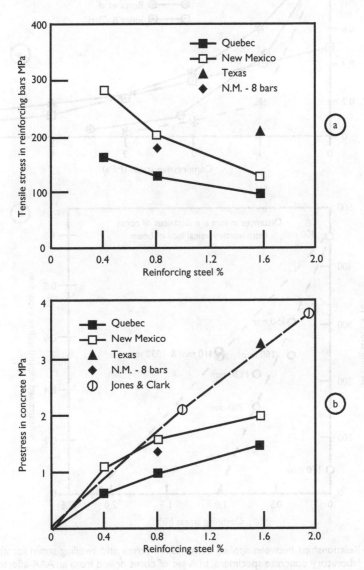

Figure 3.13 (a) Tensile stresses induced in reinforcing by expansion of concrete. (b) Compressive stresses in surrounding concrete induced by restraint imposed by reinforcing.

prevents expansion. This raises the question of how large the fully constrained swelling pressure of AAR-affected concrete is likely to be. A partial answer to this question can be found from Figure 3.12b which suggests that with 2½ to 4% of the concrete area occupied by steel, the induced concrete stress is of the order of 4 to 5 MPa, with the induced stress approximately constant as the reinforcing % increases. A second indication is given by Figure 3.14a (by the British Institution of Structural Engineers,

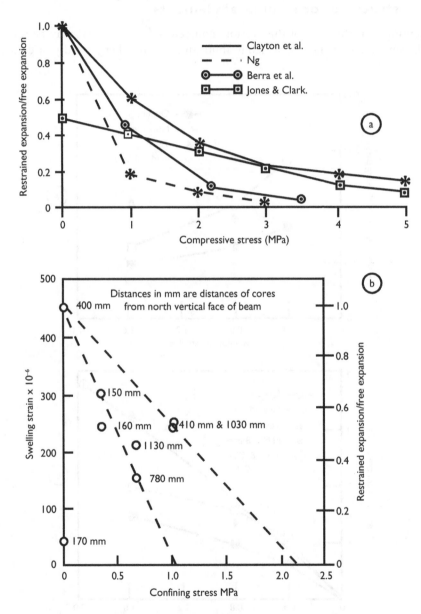

Figure 3.14 Relationships between applied compressive stress and swelling strain for a) Two sets of laboratory concrete specimens, b) A set of cores drilled from an AAR-affected beam.

1992) which shows the relationship between induced concrete stress and the ratio of restrained expansion to free expansion in tests by Clayton, et al. (1990) and Ng (1991). These results suggest that with complete restraint of swelling, the swelling pressure would be between 3 and (possibly) 5 MPa. Later measurements by Berra, et al. (2008) tend to agree with the earlier work.

The considerations set out in Section 5.2 conclude that if the relative humidity in the pores of the concrete can be reduced to and maintained at 95%, the AAR reaction will be suppressed. A relative humidity of 95% corresponds to a tensile stress in the water contained by the concrete of about 6.5 MPa. This, in turn represents the potential swelling pressure of the concrete. Because the concrete would be only partly saturated with water in this condition, the actual swelling pressure would be less than this, say 4 to 5 MPa. Hence both measurement and theory point to 4 to 5 MPa as being the upper limit to the swelling pressure induced by AAR. However, it should be noted from Figure 3.12 that all of the data were obtained from artificially accelerated laboratory tests and that the really high induced concrete stresses were related to high rates of expansion (0.5 mm/day or more). In a field situation with concrete exposed to seasonal weather, where swelling, whether or not restrained by reinforcing, develops over periods of 1 to 2 decades, the time-dependent phenomena of creep and relaxation (see Section 3.7) will come into play. Restrained expansion and induced stress in the concrete will therefore both be less.

The specimens measured in Figure 3.14a were made and tested in the laboratory with the AAR accelerated. Figure 3.14b shows a set of measured strains and the corresponding swelling pressures for the cores that were considered in Figure 3.10. The concrete was about 25 years old at the time, and much of the potential swelling had probably already occurred. These test measurements suggest a residual swelling pressure of 1 to 2 MPa. Note that 5 of the 7 specimens show a constant ratio of swelling strain to confining stress which gives confidence in their reliability.

As with Figure 3.12, it is unlikely that swelling pressures such as those shown in Figure 3.14a, derived from accelerated laboratory tests, will actually develop in real structures. Those shown in Figure 3.14b, being induced by restrained swell, residual after 25 years of restrained swell in the structure, are likely to be less than potential field values in new structures. Hence actual potential AAR swelling pressures in real structures will probably lie in the range from 2 to possibly 4 MPa. (Also see Section 5.8.1).

3.9 FRACTURING OF REINFORCING STEEL IN AAR-AFFECTED STRUCTURES

In recent years there has been a spate of reports of fractured steel stirrups and links being found in AAR-distressed structural members (e.g. Kuzume, et al., 2004, Torrii, et al., 2004). Photographs and yield stresses quoted in these reports show that the fractures occurred in cold worked, ribbed, high tensile steel links that appear to have been bent to radii approaching those of the main bars they were linking. An elementary knowledge of bending theory and the behaviour of steel under stress is required to understand why these fractures occurred.

Figure 3.15a shows (schematically) the behaviour of a mild steel bar subjected to tensile stress. (Very similar behaviour occurs under compressive stress). As the stress is increased, the steel strains elastically according to the relationship $\varepsilon = \sigma/E$, where ε is the axial strain in the bar, σ is the applied stress and E is the elastic modulus. When the yield stress is reached, the steel strains plastically at constant stress, but with

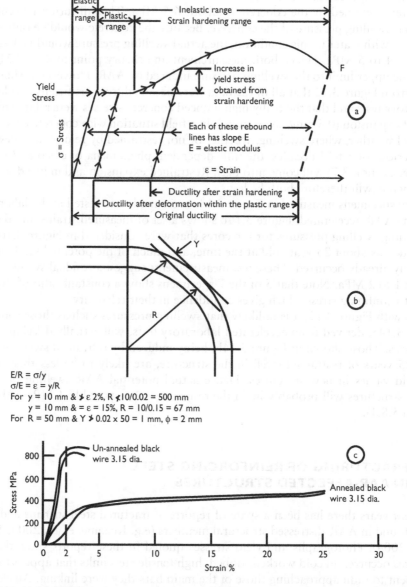

Figure 3.15 (a) Effect of strain hardening on the strength and ductility of a mild steel bar. (b) Relationship between elastic modulus E, stress in steel σ, bar radius y and bending radius R. (c) Example of stress-strain characteristics for the same steel, strain-hardened and annealed.

increasing strain, starts to strain harden, and the applied stress has to be increased to cause further strain. Finally the stress necessary to cause further straining reduces and the bar fractures at point F.

If, the stress is reduced within the plastic range, the strain will decrease along a σ vs ε line with slope E, leaving the bar permanently strained by the amount of plastic strain it has undergone. If the stress is increased again, the steel will follow the same σ/ε path as the unloading path. The same process will occur once the strain hardening range has been entered, but when the bar is reloaded, the yield stress, most importantly, will be found to have increased and the plastic strain before strain hardening will not recur. Also the strain available until failure occurs, the ductility, will have decreased. This is the process of cold working that is used to produce a high tensile steel reinforcing bar from a mild steel stock.

Figure 3.15b illustrates the relationship between the radius to which a bar is bent, and the strain to which the steel is subjected. In bending a steel bar to a radius, the yield stress must be exceeded, or else the bent bar will elastically spring back to straight. Hence, steel that is subjected to bending to a permanent radius will end up well into the strain hardened range and have a reduced ductility. This is the root of the problem with bending cold worked bars. Once bent, the bar has very little ductility at the bend and because it is bent to act as an anchorage, there will be a stress concentration at the bend. Hence the bar will fracture at the bend, with the fracture usually initiating on the outside of the bend where the permanent strain is tensile. On the inside of the bend, the permanent strain will be compressive, which will also reduce the ductility.

Figure 3.15c shows an extreme example of the effect of cold working on the properties of steel. It shows two tensile tests on a cold drawn wire which has a yield stress of 800 MPa and ductility to fracture of less than 5%. If this wire is annealed by heating to 650°C and keeping it at this temperature for about an hour, the effects of cold working are removed by recrystallization of the steel. Once the wire has cooled, as shown in Figure 3.15, its yield stress has reduced from over 800 MPa to less than 400 MPa; but its ductility has increased from less then 5% to 25% or more.

The numbers on Figure 3.15b show that it would be impossible to bend a 20 mm (2y) diameter bar of the cold drawn steel wire to a radius of less than 500 mm without fracture or cracking occurring, whereas limiting the strain to 15% with the annealed steel would allow a bending radius of 67 mm. However, if the 20 mm diameter link were to be bent to a radius of 40 mm, the permanent strain would be 25% and even with the extremely ductile annealed steel, the bend would crack and subsequently break.

Plate 3.1 shows an example of a link broken at a too sharp radius, re-photographed from a paper by Seto, et al. (2004). The arrow on the right points to a broken link bent to a sharp radius around a main bar passing across the centre of the field of view. The arrow pointing downwards points to a fracture in a weld splicing the main bar. It is obvious that both links and main bars are of cold-worked high tensile steel, and the radius of the link was made to fit the radius of the main bar. The radius was far too small, the resulting strains too large, and hence the tension failure occurred. The weld was also a poor technique to use and it probably failed because the weld metal cooled much too quickly and was embrittled by the rapid cooling, while the metal of the main bar, on either side of the weld was partly annealed by the heat of the weld. This introduced a strain incompatibility at the weld, which suffered a brittle failure. The diameters of the main bars and links were 16 and 10 mm respectively. The main steel had a yield stress of

349 MPa and the links of 378 MPa. The yield strain of the links in tension was therefore 378/30 000 = 1.26%. If the inside radius of the bends in the 10 mm diameter links was 20 mm, the outside radius would have been 30 mm and the strain on the outside of the bend would have been 5/30 = 17%, which seems (and was) too large for the fairly lightly strain hardened steel. (Also see Section 5.9 for a discussion of the repair of this failure).

3.10 THE POSSIBILITY OF BOND FAILURE IN AAR-AFFECTED REINFORCED CONCRETE STRUCTURES

As described in Sections 3.2 and 3.3 (Figures 3.1a and b), AAR expansion results in cracking of the covercrete, with a preferential tendency for cracking parallel to the surface of the concrete. Cracks can, and do penetrate to below the level of the steel. The British Institution of Structural Engineers (1992) drew attention to the distinct possibility of bond failure, especially in bars that are not restrained by links or where substantial cover has not been provided. Based on work carried out at the British Transportation and Road Research Laboratory and by Cairns and Jones (1996), the following limitations on bond stress in concrete likely to be affected by AAR are suggested:

For plain bars, the characteristic bond strength, f_{bs}, should be taken as:

$$f_{bs} = \beta f_t \tag{3.3}$$

where f_t is the characteristic splitting or indirect tensile strength of the AAR-affected concrete (see Section 3.4 where f_t has been termed σ_{it}) and

β is a coefficient, generally taken as 0.65, but reduced to 0.45 for a top cast bar and 0.33 for a corner and top cast bar.

For ribbed bars

$$f_{bs} = \propto (0.5 + c/d)f_t \tag{3.4}$$

where c = cover, b = bar diameter, and

$\propto$ is a coefficient, generally taken as 0.6, but 0.4 for a top cast bar and 0.3 for a corner and top cast bar.

In all cases, cover should be at least three bar diameters, and the spacing of bars should not be less than five diameters (Cairns and Jones, 1996, American Concrete Institute, 1991).

Ahmed, et al. (1999) showed that AAR reduces the strength of lap splices in bars carrying tension, showing that in laboratory conditions, a beam with lap lengths of 46 diameters in sound concrete will perform satisfactorily, whereas if the concrete is affected by ASR, a lap of 54 diameters is likely to fail.

The literature does not seem to have reported any notable failures in AAR-damaged structures caused by bond failure. The Institution of Structural Engineers (1992) cautions that delamination in the plane of the reinforcing may occur in slabs as a result of AAR, effectively dividing the slab into 3 layers if there are no links tying the top and bottom reinforcing together. This could cause a loss in strength of up to 30%. They recommend coring of AAR-affected slabs to check for possible delamination.

Bond failures do not seem to be very common, but can have disastrous consequences, as illustrated by the failure of a very large cement storage silo that occurred in 2003. This was not as a consequence of AAR, but is of interest as it very well illustrates the signs of an impending bond failure, as well as the consequences (Blight, 2006). After 3 years of service, the multi-cell silo which had an inner cylindrical cell surrounded by 4 outer segmental cells, was emptied and cleaned for inspection. It was found that the inner cylindrical wall showed closely spaced vertical cracks, while the outer cylindrical wall of the cell showed similar cracking, although not as severe. Figure 3.16 has been included to show the cracking that was mapped on these two opposite walls, which is typical of actual or impending bond failure. In this case,

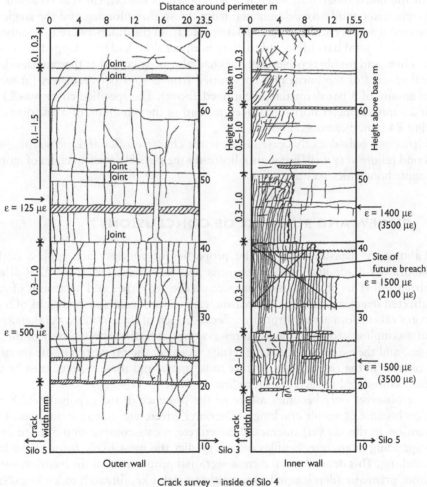

Figure 3.16 Signs of imminent bond failure as shown by the intense parallel cracking of the inner and outer walls of a cement storage silo.

the silo was 59 m high and 25 m in outer diameter, so this was a bond failure on an enormous scale.

Before an investigation into the causes and possible consequences of the cracking could be completed, it was decided, purely for commercial reasons, to refill this particular outer cell with 7000 tonnes of cement. Filling was completed and the full silo cell was left standing undisturbed for 16 hours, whereafter emptying was started. Almost immediately, the outer wall split open, with a strip tearing away, from top to bottom. It was then found that the inner wall had also breached allowing cement to pour into the central silo cell that had been almost empty at the time.

The only possible course of action was to demolish the silo, working from the top down and salvaging the contained cement as the work progressed. (The value of the cement was more than the cost of rebuilding the silo). Plate 3.2 shows the condition of the failed inner wall when it was exposed as the cement was removed. The covercrete had completely delaminated from the steel reinforcing and the single mat of horizontal and vertical steel had delaminated from the inner concrete. Numbers of adjacent horizontal bars (designed to carry hoop tension) had been lapped in the same vertical line, and numbers of vertical bars had also been lapped at the same level. The vertical spacing of the horizontal bars varied from 3 to 4 bar diameters (instead of the minimum of 5 bar diameters, mentioned above). The specified cover was 25 mm to the 28 mm diameter horizontal steel (instead of the minimum of 3 bar diameters, equal to 84 mm, mentioned above).

Hence not only does this case illustrate the characteristics and the warning signs of a bond failure very well, but it also illustrates the possible consequences of ignoring the simple bond rules set out earlier.

3.11   REVIEW AND SUMMARY OF CONCLUSIONS

- Laboratory measurements of the properties of concrete are applied both to laboratory-made and tested specimens and to cores taken from AAR-affected structures. This book has concentrated on the properties and behaviour of AAR-affected structures, and for this reason has mainly considered the results of testing cores taken from actual structures. Because each of these cores was taken as part of a sampling program, the usual reservations as to the representivity of the samples, and the extent to which the results represent the properties of the remaining concrete in the rest of the structure must apply, and these aspects must be carefully considered in each case. (See Section 3.1).
- The observer only becomes aware of the presence of the expansive AAR reaction because of visible cracking of the outer covercrete. A simple analysis of how tension in the (as yet) uncracked covercrete resists compression applied by the expanding heartcrete, enables one to predict the most likely form of the initial cracking. This depends on the cross-sectional proportions of the member, varying from prismatic (depth equal to breadth) to slab-like, (breadth much in excess of depth). The analysis shows that cracking is always most likely to occur longitudinally, and in slab-like members, longitudinal cracks are likely to occur parallel to the slab surface (i.e. cracking will tend to delaminate the covercrete). The analysis

is also valid for wall-like elements (depth much in excess of breadth) because a wall is simply a vertical slab. (See Section 3.2).

- In an actual structure, as opposed to a laboratory specimen, the AAR process occurs very slowly, the usual period between completion of construction and diagnosis of AAR being 10 to 15 years. Exceptions appear to have been the intake tower of the Itezhitezhi dam, in which the effects of concrete swelling were noticed after only 4 years (Thaulow, 1983) and the underground concrete plug described in Section 4.10.5. In the latter case, the development of AAR swelling was clearly accelerated by the high ambient temperature (38°C) and the constant 100% relative humidity environment. Also, in an actual structure, expansion is restrained by the presence of reinforcing, by adjacent structural elements, etc. The net effect is that strains in concrete taken from the field are generally very much less than strains measured in laboratory specimens where AAR has been artificially accelerated. (See Section 3.3).

- Laboratory testing, as with any activity depending on the engineering measurement of load and deflection, has to be performed very carefully, with due cognizance of the effects of extraneous stresses introduced by end conditions, etc. In general, the result of a test can only be as good as the calibration of the measuring instruments, the avoidance of extraneous boundary effects and the care of the testing technician. (See Section 3.4).

- It often seems to be assumed that once concrete has failed in compression, it has no further reserve of strength. This is not, however, the case. As long as a disintegrated concrete carries compressive triaxial stresses, it has a surprisingly large reserve of strength. (See Section 3.5)

- The ratio of the elastic modulus E of AAR-affected concrete to its compressive strength σ_c, based on a wide range of published results, varies from

$E/\sigma_c = 300$ to 1000

The ratio of σ_c to the indirect tensile strength, σ_{it} for AAR-affected concrete varies from

$\sigma_c/\sigma_{it} = 11$ to 20

For the direct tensile strength of AAR-affected concrete σ_t, the ratio

$\sigma_c/\sigma_t = 20$ to 60

Hence the direct tensile strength falls well below the indirect tensile strength.

- Values of Poisson's ratio range between 0.19 and 0.29 and appear to be independent of the strength of the concrete. (See Section 3.6).

- Laboratory creep tests on cores taken from an AAR-affected structure were found to conform reasonably well with equation 3.2:

$$\varepsilon(t) = \sigma\left(\frac{1}{E} + \frac{t}{\eta}\right) \tag{3.2}$$

with E varying from 31.5 GPa for apparently undamaged concrete (A) to 12.6 GPa for concrete showing moderate to severe damage (C), and η, the structural viscosity, varying from 24.7 GPa years for (A) to 5.6 GPa years for (C). (See Section 3.7). It was noted that creep in actual structures in situ appears to be very much less than that measured under uniaxial loading in laboratory tests.

- The most common form of restraint on AAR-expansion is that offered by embedded reinforcing, with restraint provided by adjacent or abutting structures being very much less common. Most of the research on compressive stresses induced in concrete by restrained expansion and conversely, tensile stresses induced in reinforcing steel, has been done on laboratory specimens with the development of AAR being accelerated in various ways. These tests inevitably result in exaggerated results, with compressive stresses induced in the concrete of up to 14 MPa and corresponding steel stresses of up to 140 MPa. However, measurements on cores taken from AAR-affected structures as well as other theoretical considerations point to a range of maximum possible swelling pressures of 4 to 5 MPa with induced tension stresses in reinforcing steel of up to 40 to 50 MPa. (See Section 3.8).

- In recent years, fractured reinforcing, usually high tensile, ribbed links wrapping around main reinforcing have been discovered on investigation of AAR damage. In fact, the problem has nothing to do with AAR, but results from poor detailing of bars, which calls for the bending of high tensile steel links to unrealistically sharp radii. (See Section 3.9).

- Section 3.2 showed that there will be a tendency in slab or wall-like reinforced concrete members subjected to AAR, to suffer delamination-type cracking parallel to, or within the plane of the reinforcing. Special attention should be paid to the possibility of bond failure by ensuring an adequate cover thickness (not less than 3 bar diameters) and adequate bar spacing (not less than 5 bar diameters). Lap splices are particularly vulnerable to bond failure, and straight lap lengths should exceed 60 diameters. Top and bottom rainforcing mats in slabs and opposite mats in walls should be linked together at regular intervals to further reduce the possibility of bond failure.

REFERENCES

Ahmed, TMA, Burley, E & Rigden, SS 1999, 'Effects of alkali-silica reaction on the tensile bond strength of reinforcement in concrete tested under static and fatigue loading', *ACI Material Journal*, no. 96 – M52, pp. 419–428.

Alexander, MG, Blight, GE & Lampacher, GJ 1992, 'Pre-demolition tests on structural concrete damaged by AAR', *9th Int. Conf. on AAR in Concrete*, London, U.K., vol. 1, pp. 1–10.

American Concrete Institute 1991, *Commentary on standard practice for design and construction of concrete silos and stacking tubes for storing granular materials*, ACI 313R – 91, The Institute, Farmington Hills, Missouri, U.S.A.

Berra, M, Faggiani, G, Mangialardi, T & Paolini, AE 2008, 'Influence of stress constraint on the expansive behaviour of concrete affected by ASR', *13th Int. Conf. on AAR in Concrete*, Trondheim, Norway, p. 10. (No page numbers, available on CD).

Blight, GE 2007, 'The delayed collapse of a 42 000 ton capacity RC cement storage silo', *Magazine of Concrete Research*, vol. 59, no. 5, pp. 329–340.

Blight, GE, McIver, JR, Schutte, WK & Rimmer, R 1981, 'The effects of alkali-aggregate reaction on reinforced concrete structures made with Witwatersrand quartzite aggregate', *5th Int. Conf. on AAR in Concrete*, Cape Town, South Africa, Paper S252, p. 12.

Blight, GE & Alexander, MG 1988, 'Evaluating reinforced concrete structures affected by alkali aggregate reaction', *Int. Symp. on Re-evaluation of Concrete Structures*, Copenhagen, Denmark, pp. 309–317.

British Concrete Society 1976, *Concrete core testing for strength*, British Concrete Society, Technical Report No. 11, London, U.K.

British Institution of Structural Engineers 1992, *Structural effects of alkali-silica reaction*. The Institution, London, U.K.

Cairns, J & Jones, K 1996, 'An evaluation of the bond-splitting action of ribbed bars', *ACI Materials Journal*, vol. 93, no. 1, pp. 10–19.

Clayton, N, Currie, RJ & Moss, RM 1990, 'The effects of alkali-silica reaction on the strength of prestressed concrete beams', *The Structural Engineer*, vol. 68, no. 15, pp. 287–292.

Hasparyk, NP, Lopes, AN de M, Tavares Cavalcanti, AJC & Silveira, JFA 2004, 'Deterioration index and properties of concretes from Paulo Afonso power plants, Brazil, affected by the alkali-aggregate reaction', *12th Int. Conf. on AAR in Concrete*, Beijing, China, pp. 898–905.

Jones, AEK & Clark, LA 1996, 'A review of the Institution of Structural Engineers', "Structural effects of alkali-silica reaction (1992)": *10th Int. Conf. on AAR in Concrete*, Melbourne, Australia, pp. 394–401.

Jones, R & Gatfield, EN 1955, *Testing Concrete by an ultrasonic pulse technique*, Road Research Technical Paper No. 34, HMSO, U.K.

Kuzume, K, Matsumoto, S, Minami, T & Miyagawa, T 2004, 'Experimental study on breaking down of steel bars in concrete structures affected by alkali-silica reaction', *12th Int. Conf. on AAR in Concrete*, Beijing, China, pp. 1162–1168.

Ng, KF 1990, *Effect of alkali-silica reaction on punching shear capacity of reinforced concrete slabs*. PhD. Thesis, University of Birmingham, U.K.

Nomura, N, Kakio, T, Matsuda, Y & Nishibayashi, S 2004, 'Investigation and repair process of fractured reinforcements due to ASR', *12th Int. Conf. on AAR in Concrete*, Beijing, China, pp. 1271–1276.

Oberholster, RE 1989, 'Alkali-aggregate reaction in South Africa: some recent developments in research', *8th Int. Conf. on AAR in Concrete*, Kyoto, Japan, pp. 77–82.

Ono, K 1989, 'Assessment and repair of damaged concrete structure', *8th Int. Conf. on AAR in Concrete*, Kyoto, Japan, pp. 647–658.

Siemes, T & Visser, J 2000, 'Low tensile strength in older concrete structures with alkali-silica reaction', *11th Int. Conf. on AAR in Concrete*, Quebec City, Canada, pp. 1029–1038.

Sims, GP & Evans, DE 1988, 'Alkali-silica reaction: Kamburu spillway, Kenya, case history', *Proc. Instn. Civ. Engrs.*, Part 1, vol. 84, pp. 1213–1235.

Smaoui, N, Berube, M-A, Fournier, B & Bissenette, B 2004, 'Stresses induced by ASR in reinforced concrete incorporating various aggregates', *12th Int. Conf. on AAR in Concrete*, Beijing, China, pp. 1191–1201.

Thaulow, N 1983, 'Alkali-silica reaction in the Itezhitezhi dam project, Zambia', *6th Int. conf. on AAR in Concrete*, Copenhagen, Denmark, pp. 471–477.

Torii, K, Sannoh, C, Kubo, Y & Ohashi, Y 2004, 'Serious damages of ASR affected RC bridge piers and their strengthening techniques', *12th Int. Conf. on AAR in Concrete*, Beijing, China, pp. 1283–1288.

Wen, H-X, Wang, Y-Q & Balendran, RV 2000, 'Damage mechanics model for AAR-affected concrete', *11th Int. Conf. on AAR in Concrete*, Quebec City, Canada, pp. 1039–1048.

PLATES

Plate 3.1 Fractured steel reinforcing exposed in reinforced concrete member damaged by AAR.

Plate 3.2 Large scale bond failure accompanying delamination of covercrete in reinforced concrete silo wall.

Chapter 4

Assessment of risk of structural failure based on the results of laboratory or field tests

4.1 INTRODUCTION, DEFINITIONS AND EXAMPLES

Failure of a structure has occurred when it no longer meets its intended design and performance requirements. These may include aesthetic considerations. If a reinforced concrete structure has developed signs of AAR it will usually have failed aesthetically in the sense that its appearance will have deteriorated and its condition may appear alarming, but it may remain perfectly serviceable from strength and/or deformation aspects. A reinforced concrete structure may also become unserviceable as a result of deformation or dimensional change or exposure of the reinforcing steel to corrosion, but retain adequate strength for the time being.

The probability of failure may be defined as the probability that the loading applied to the structure will exceed its strength, or the strength of one of its components. However, here it will also be taken to mean the probability that a structure will become unserviceable as a result of deformation, dimensional change or cracking resulting from AAR.

For any structure there are several possible modes of failure that can result from many potential causes. Each mode may have an associated chain of consequences. The probability of failure could also be defined as the probability that a given mode of failure will occur and will result in the associated chain of consequences. The financial risk (or simply the risk) is defined as the product of the sum of the costs of rectifying the consequences (e.g. repair or rebuilding) plus the financial loss caused by the consequences (e.g. the cost of loss of production, penalties or law-suits arising out of failure to meet contractual obligations, etc.) and the probability of failure, i.e.:

Financial risk = Probability of failure × Cost associated with failure   (4.1)

(Also see Ho, et al., 2000.)

The following is an illustrative hypothetical example of risk which is, however, based on two real case histories:

A reinforced concrete silo used at a colliery to load trains with coal for export, was found to be showing signs of damage from AAR. Silos are interesting subjects for study, as the contents of a silo are often worth three to four times the value of the structure.

Several questions arose, of which the following are examples:

- What is the current extent of the damage?
- Is it dangerous for people to work near or on the structure?
- Will the AAR progress, resulting in an ever-increasing extent of damage? At what rate might it progress?
- Will the AAR result in a need to repair the silo extensively, and if so, how soon?
- If it is not repaired, will it fail structurally, and how will it fail?
- Without intervention, what is its remaining service life? To what extent will extensive intervention cure the problem?
- What are likely modes of failure, and what will be their likely consequences?

As the silo served a rail-line with the traction provided by overhead high voltage power lines, and as extensive repair would require the erection of scaffolding around the silo, the export of coal from the mine would need to be suspended during any, except very limited and localized, repair operations. This would cause the mine to lose profit at a rate of the equivalent of 50 000 US dollars per day. Furthermore the mine would have to pay penalties for non-performance of contractual obligations of the equivalent of 20 000 US dollars per day.

Obviously, in a case like this, the financial risk of continuing to operate the silo may be so large that there may be less risk in building a duplicate silo in which every precaution is taken to ensure that AAR will not recur. Once the duplicate facility has been completed, the switch-over from the damaged silo to the new one can be made with little or no loss of export coal quantities.

In one of the real case histories, the silo and its ancillaries were duplicated and taken into service and the damaged silo was then demolished. In the other case, it appeared preferable to accept the financial risk, large as it was, and the damaged silo continued in operation, with regular and careful monitoring of its condition. The danger of this route is that with changes in the mine personnel, the monitoring may be forgotten or neglected in a year or two's time because no change in the silo's condition is observed. If the condition of the silo does indeed deteriorate, the ensuing failure may be a surprise and a career-threatening shock to all concerned. See Plate 4.1 for an example of a silo being kept in operation, with twice-daily coal loadouts, while being investigated for possible structural inadequacy.

4.2   AN ACCEPTABLE PROBABILITY OF FAILURE

In considering the safety of a structure, the probability of failure is paramount and cost considerations should be secondary. In the case of a possible structural collapse, what is an acceptable probability of failure? In other words, what probability of death or injury caused by collapse of a structure is considered acceptable by the public?

Figure 4.1 displays world fatality statistics for accidents or failures in a number of industrial activities. Statistics are in the form of annual probability or frequency of failure (F) versus the number (N) of associated fatalities and the monetary cost of rectifying such a failure. Note that in commercial aviation, the probability of an accident

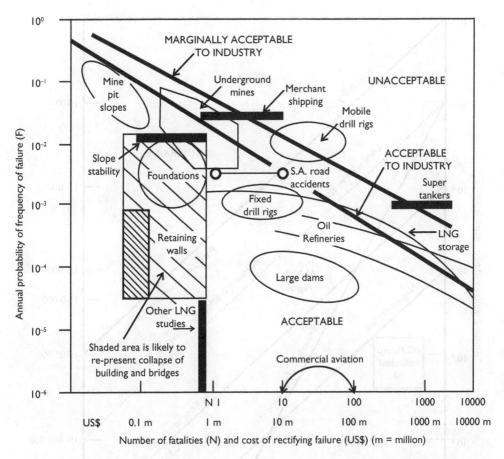

Figure 4.1 Industrial data on probability of failure and associated fatalities for various industrial activities (after Whitman, 1987, with additions from Blockley, 1995 and authors).

is about 1 in a million, and the likely associated death-toll of an accident, between 10 and 100. This is obviously quite acceptable to the burgeoning flying public. The probability of a merchant ship having an accident is about 1 in 20, and the associated likely death toll is up to 10. Road accident statistics for South Africa fall just on the "acceptable" line with an annual probability of being involved in an accident of 1 in 2000 and a likely death toll per accident of 1 to 10. No statistics are shown for residential buildings or bridge structures, but the range of F is likely to be similar to that for foundations and retaining walls, and the range of N could be 1 to 100, per accident.

The so-called F-N diagram in Figure 4.1 records actual statistics. Figure 4.2 shows limiting lines on an F-N diagram that summarizes institutional opinion on acceptable positions for limiting F-N lines. Rather curiously, all lines are based on one point, that labeled "BC Hydro individual risk" which, with co-ordinates $F = 10^{-4}$ (or 1 in 10 000)

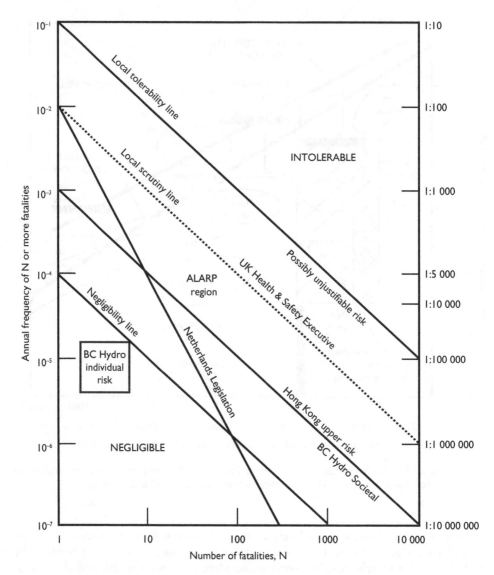

Figure 4.2 Guide to what are publicly acceptable probabilities of failure, (after Canadian Dam Safety Association, 1995 and US National Research Council, 1989, quoted in Blockley, D. (1995)).

and $N = 1$, represents the probability of death by natural causes of any member of the safest population group in North America, 10 to 14 year-olds. Any F-N combination falling below this line represents a "negligible risk". At the other extreme, a probability of death of 1 in 10 for one person or of 1 in 1000 for a group of 100 people is taken to be the upper limit to acceptability. The region between these bounding lines, is labeled ALARP (i.e., keep it As Low As Reasonably Possible) and is also an acceptable region. The limiting value of 1 in 10 000, as a probability for a single death that is acceptable to society, is a useful benchmark to remember. If the probability of structural collapse is assessed as more than an acceptable value, the structure should be taken out of service and/or repairs undertaken urgently, possibly including such measures as reducing or restricting the live loading on the structure, restricting access, or introducing temporary supports.

In the case of failure as a loss of serviceability but with negligible danger of structural failure, a larger probability of failure could be acceptable, depending on the financial risk. For example, if there is no probability of death or injury, more than 1 in 10 000 or even 1 in 1 000 may, under the prevailing circumstances, be considered to be an acceptable risk. The decision now becomes financial rather than one of public safety.

Assessing the probability of failure is the first step in assessing the financial risk. A simple but well-established method of doing this is given in what follows.

PART I Statistical considerations

4.3 STATISTICAL CALCULATION OF THE PROBABILITY OF FAILURE

Engineering design aims to ensure that there is a safe margin between stresses caused by the applied physical and environmental loads (or the structural "Demand") and the available strength of the structure or structural component (the structural "Capacity"). Thus, for failure not to occur, the demand D must always be less than the capacity C. However, neither C nor D are single valued, both will have statistical distributions. The situation is illustrated in Figure 4.3 which shows the "demand/capacity" model, with both capacity C and demand D represented by statistical distributions (Rackwitz and Fiessler, 1988, Hohenbichler and Rackwitz 1988, Haurylkjiewicz, 1979). Conventionally, the factor of safety would be represented by

$$FS = C^1/D^1 \qquad (4.2)$$

Where C^1 and D^1 are the single values of C and D judged by the designer to be "representative", i.e. a low enough capacity and a high enough demand to give a conservative value of FS.

In Figure 4.3, $\overline{C}$ and $\overline{D}$ are the mean values of C and D, and their ratio

$$CFS = \overline{C}/\overline{D} \qquad (4.3)$$

is called the "central factor of safety". CFS is usually greater than FS.

The difference between C and D is the safety margin and failure (or damage) will occur if (C − D) is less than or equal to zero. However, (C − D) is itself a random variable and hence there is a "probability of failure" (p_f) which is the probability that the safety margin will be less than zero. This probability is represented by the shaded area

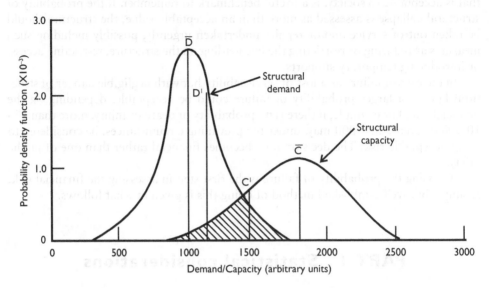

Figure 4.3 Demand/capacity model illustrating equations (6.2) to (6.5).

of overlap in Figure 4.3. The smaller the area of overlap, the less will be the probability of failure. Mathematically

$$p_f = \text{Probability } (C < D) \tag{4.4}$$

As an example, if C and D have statistically normal (i.e. Gaussian) distributions

$$p_f = 1 - F\left[\left(\overline{C} - \overline{D}\right)\big/\left(\sigma_C^2 + \sigma_D^2\right)^{1/2}\right] \tag{4.5}$$

Where F (for the normal distribution) is the function tabulated in Table 4.1 and σ_C and σ_D are the standard deviations of C and D, respectively.

In some cases C and D cannot be taken to be normally distributed in which case the appropriate value of F for the actual distribution must be used. For the purposes of this section, all statistical variables will be taken to be normally distributed.

The main problem of applying equation (4.5) in practice is that it is seldom that sufficient test data are available to enable C and D to be well defined in terms of $\overline{C}, \overline{D}, \sigma_C, \sigma_D$. In particular, D depends on the actual applied loading which may be very different from the design loading, according to the relevant design code. C, in turn, depends on the actual strength distributions of the material or materials of which the structure is constructed. In the case of a reinforced concrete structure affected by AAR, the concrete properties can be expected to deteriorate progressively, the current strength distributions (compressive, tensile, shear) can be estimated by various methods, but these estimates give only the current values. The decline of these strengths with time (if there will be a decline) is obviously also of great importance, even if current values are satisfactory.

Table 4.1 The cumulative distribution function F(u) for the standard normal distribution.

u	0.00	0.01	0.02	0.03	0.04	0.05	0.06	0.07	0.08	0.09
0.0	0.500	0.503	0.507	0.511	0.515	0.519	0.523	0.527	0.531	0.535
0.1	0.539	0.543	0.547	0.551	0.555	0.559	0.563	0.567	0.571	0.575
0.2	0.579	0.583	0.587	0.590	0.594	0.598	0.602	0.606	0.610	0.614
0.3	0.617	0.621	0.625	0.629	0.633	0.636	0.640	0.644	0.648	0.651
0.4	0.655	0.659	0.662	0.666	0.670	0.673	0.677	0.680	0.684	0.686
0.5	0.691	0.694	0.608	0.701	0.705	0.708	0.712	0.715	0.719	0.722
0.6	0.725	0.729	0.732	0.735	0.738	0.742	0.745	0.748	0.751	0.754
0.7	0.758	0.761	0.764	0.767	0.770	0.773	0.776	0.779	0.782	0.785
0.8	0.788	0.791	0.793	0.796	0.799	0.802	0.805	0.807	0.810	0.813
0.9	0.815	0.818	0.821	0.823	0.826	0.828	0.831	0.833	0.836	0.838
1.0	0.841	0.843	0.846	0.848	0.850	0.853	0.855	0.857	0.859	0.862
1.1	0.864	0.866	0.868	0.870	0.872	0.874	0.876	0.879	0.881	0.882
1.2	0.884	0.886	0.888	0.890	0.892	0.894	0.896	0.897	0.899	0.901
1.3	0.903	0.904	0.906	0.908	0.909	0.911	0.913	0.914	0.916	0.917
1.4	0.919	0.920	0.922	0.923	0.925	0.926	0.927	0.929	0.930	0.931
1.5	0.933	0.934	0.935	0.936	0.938	0.939	0.940	0.941	0.942	0.944
1.6	0.945	0.946	0.947	0.948	0.949	0.950	0.951	0.952	0.953	0.954
1.7	0.955	0.956	0.957	0.958	0.959	0.959	0.960	0.961	0.962	0.963
1.8	0.964	0.964	0.965	0.966	0.967	0.967	0.968	0.969	0.969	0.970
1.9	0.971	0.971	0.972	0.973	0.973	0.974	0.975	0.975	0.976	0.976
2.0	0.977	0.977	0.978	0.978	0.979	0.979	0.980	0.980	0.981	0.981

Values of $1 - F(u)$ for $u > 2$

u	2.32	3.09	3.72	4.27	4.75	5.20	5.61	6.00	6.36	6.71
	10^{-2}	10^{-3}	10^{-4}	10^{-5}	10^{-6}	10^{-7}	10^{-8}	10^{-9}	10^{-10}	10^{-11}

4.4 ASSESSING DEMAND D AND CAPACITY C

In order to assess a probability of failure, the demand D and capacity C for the structure must be assessed.

4.4.1 Assessing the demand D

As outlined above, the demand on a structure (D) represents the spectrum or distribution of loads and load-effects to which the structure is expected to be subjected during its working life. These are usually specified by the design code of practice to which the structure is designed. For many structures, assessment of actual, rather than design loads may be difficult. By their very nature, loading codes are very conservative and where loads are ill-defined, the loads they call for may be unrealistic. For example, an elevated motorway structure in South Africa had been designed for British Standard (BS 153, British Standards Institution, 1954) type HA loading. For this particular structure, HA loading amounted to uniformly distributed loads of 33 kN/m and 7 kN/m plus concentrated live loads of 1 223 kN and 310 kN over two traffic lanes each 3.66 m wide and a third lane 2.33 m wide, respectively. When it came to planning a full-scale loading test on the structure, it was found to be physically impossible to apply full HA loading using road trucks overloaded to 10% above their legal capacity (with the special permission of the traffic police) and positioned as closely nose-to-tail and side-by-side as was possible (see Plate 4.12, where two more lanes of trucks still have to be positioned). With this arrangement only 84% of HA loading could be achieved. (It should be noted that this was the required design load, not a factored proof load.) Six years later when the structure was re-tested a (then new) South African loading code (TMH 7) (South African National Institute for Transport and Road Research, 1981) had been introduced. Although the TMH 7 or NA loading is not directly comparable with the HA loading it amounts to about 81% of HA loading and it was physically possible to fit this loading on the roadway. However, this does not mean that the NA loading is realistic, as the probability of such a load actually occurring would be extremely low.

In another case relating to the same elevated motorway, a column supporting the motorway was found to have suffered damage from AAR. As part of the investigation of consequences, long gauge length strain gauges were affixed longitudinally to the column and strains were recorded during peak hour traffic on the motorway (Figure 4.4). The maximum dynamic stress recorded was only 0.5 MPa, whereas the dead-load stress was of the order of 15 MPa. The low value of the live loading arose because, under real traffic conditions, not the impossibly closely spaced condition assumed by the loading code, heavy vehicles are widely spaced longitudinally, and it is seldom that more than one heavy vehicle is present on any one span, or more than one traffic lane on any one span at any time.

Realistic values for D would have been 15 to 16 MPa instead of the much higher (dead + live) load that would be demanded by HA or TMH 7 highway loading.

4.4.2 Assessing the capacity C

The capacity of a structure is its probable available strength, i.e. in this case, the strength, reduced by the damage caused to the concrete by the AAR. To assess the capacity, an ultimate load analysis must be carried out taking cognizance of the reductions in compressive strength, shear strength, bond strength and modulus of elasticity, as at the time of the investigation and as predicted for future times. The resultant strength, expressed as a failure or collapse load would represent the capacity.

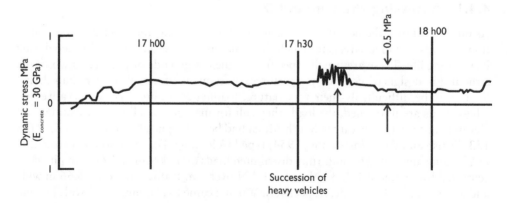

Figure 4.4 Dynamic strains recorded in column under rush hour traffic loading.

If the structure is actively deteriorating with advancing time, as the AAR progresses, the capacity distribution (Figure 4.3) will progressively move to the left, as $\bar{C}$ decreases and σ_C (possibly) increases. There is also a possibility that increasing demand on the structure will arise through changes, for example, in vehicle type and loading, as well as increasing traffic congestion. Hence the demand distribution (D) may move to the right. Hence the area of overlap of the C and D distributions in Figure 4.3 will increase and so will the probability of failure. These are aspects that need to be investigated and their effects considered.

Because available data are usually limited, the following assumptions are sometimes made in order to apply equation (4.5):

Assumption A:

$$\bar{C} = \frac{1}{2}(C_{min} + C_{max}) \tag{4.6a}$$

$$\bar{D} = \frac{1}{2}(D_{min} + D_{max}) \tag{4.6b}$$

Assumption B:

$$\sigma_C = \pm\frac{1}{2}(C_{max} - C_{min}) \tag{4.7a}$$

$$\sigma_D = \pm\frac{1}{2}(D_{max} - D_{min}) \tag{4.7b}$$

If one is considering a failure by loss of serviceability, e.g. by excessive distortion of the structure, a similar approach to the above will be taken, assessing the demand in terms of permissible deflection, crack width or loss of clearance as the case may be. The capacity of C will be assessed as the probable distortion, taking account of the effects of damage caused by AAR to the concrete at the time of the assessment.

4.5   A SIMPLE EXAMPLE OF CALCULATING p_f

Referring to the column supporting a motorway that is described above, tests on cores taken from the column gave compressive strengths such that C ranged from 28 to 50 MPa. Using the simplifying assumptions (6.6) and (6.7) and assuming that the capacity of the axially loaded column was represented by the strength of concrete cores taken from the column:

$\bar{C} = 39$ MPa, $\bar{D} = 15.5$ MPa

$\sigma_C = \pm 11$ MPa, $\sigma_D = \pm 1$ MPa

$(\bar{C} - \bar{D})/(\sigma_C^2 + \sigma_D^2)^{1/2} = 2.12$

and $p_f = 1 - 0.99 = 0.01$ or 1 in 100.

In terms of section 4.2 this was clearly too high a probability, to be acceptable. Thus the fact that live load stresses on the column were shown to be small (Figure 4.4) did not mean that the column was in a safe condition. However, better statistical information on capacity might decrease the calculated probability of failure.

The strengths actually measured on eight cores were 28, 32, 33, 38, 40, 50, 50 and 44 MPa.

The corresponding mean and standard deviation based on 8 cores would be:

$$\bar{C} = 39.38 \text{ MPa}, \sigma_C = \pm 7.95 \text{ MPa}$$

If only 4 cores had been taken, with strengths of 28, 33, 40, and 50 MPa:

$$\bar{C} = 39.05 \text{ MPa}, \sigma_C = \pm 8.04 \text{ MPa}$$

Hence in this particular example, doubling the number of cores from a minimum of 4 to 8 would have made very little difference to the estimates of $\bar{C}$ and σ_C.

Recalculating the probability of failure using the improved values for $\bar{C}$ and σ_C of $\bar{C} = 39$ MPa and $\sigma_C = \pm 8$ MPa, however makes a change in the calculated value of p_f, lowering it from 0.01 to just more than 0.002, i.e. from 1 in 100 to 1 in 500. This is still not acceptable, but much closer to acceptability than was the rough estimate. Hence good data on C and D are essential for accurately estimating the probability of failure. In this case, knowing the demand $\bar{D}$ more accurately e.g. $\sigma_D = \pm 0.1$ MPa will have little effect of p_f as its value is mainly dependent on σ_C.

The column clearly had to be strengthened, or at the very least, the inputs of C and D needed to be verified or refined.

In this case, as most of the load consists of dead load, with a very small contribution from live load, there is little doubt that the load or demand was known fairly accurately. A greater uncertainty attached to the strength or capacity of the column. This could easily, and at reasonable cost, be strengthened, for example, by sleeving it.

4.6 CONCLUSIONS ON STATISTICAL ASSESSMENT OF RISK

The technique of assessing the probability and risk of failure for a structure has been available for many years. Acceptable probabilities of failure have also been widely discussed and agreed on. The concept and techniques are simple and can easily be applied to structures that have deteriorated as a result of AAR.

The results of an investigation of the probability and risk of failure may bring peace of mind to the owners of the structure. Alternatively, it may spur them on to undertake very necessary repair measures.

PART 2 Full-scale test loading

4.7 FULL-SCALE TEST LOADING AS A MEANS OF ASSESSING RISK

Full-scale load testing is probably the ultimate measure of safety and serviceability of an AAR-deteriorated (or any other) structure. If the structure behaves predictably and in accordance with the design requirements, and if strains and deformations are predictable, recoverable, and within design limits, there can be little doubt that structural adequacy, up to the time of the test, has been maintained and demonstrated. Such testing is, however, extremely expensive. In many cases it may be sufficient to instrument the structure, or part of the structure, and observe its behaviour under service loading. An example of this approach was given in Section 4.4.1 (and is illustrated in Figure 4.4). The literature records that several complete structures damaged by AAR have been subjected to full-scale load testing during the past 30 years. Eleven examples of full scale test loadings on complete or part structures and their results will be described in Sections 4.10 to 4.12. These have been selected to illustrate the possible effects of AAR on different types of structure and in various climatic conditions. Before describing these tests and their results, Section 4.8 will describe the typical instruments used to measure deflection, strain and rotation and Section 4.9 will give guidance on planning an in situ load test.

Test loadings could be classified as "special or once or twice off", in that only one or two test loadings are undertaken, followed by monitoring of superficial features, e.g. crack width and length, to give assurance that the structure is not progressively deteriorating. The other type of test loading is the routine loading in which a test loading is carried out annually or bi-ennially (every two years) to ensure that its satisfactory structural capacity is being maintained. These are particularly valuable in demonstrating the behaviour of AAR-affected structures over a long time period. Examples of both types of loading will be given, as well as examples of test loading carried out on smaller structural components (in this case prestressed concrete railway sleepers) or parts of a larger structure cut out, transported to a laboratory and tested in well controlled laboratory conditions.

4.8 INSTRUMENTS USED FOR MEASUREMENTS IN LABORATORY AND IN SITU LOAD TESTING

This section will describe the principles, instruments and methods used to measure movements, deflections and strains of structures and structural components under test loading. Qualitative observations and investigations used for diagnostic purposes were dealt with in Chapter 2. The application of the instruments in laboratory testing was described in Chapter 3.

The measurements usually required, include strains in both concrete and reinforcing steel, displacements, e.g. bending deflection or horizontal sway, and joint rotation. Strains in concrete need to be known both in magnitude and direction and principal strains and directions must be determinable in some cases. The methods most commonly used during the past 50 years (and used in the case histories described in Sections 4.10, 4.11 and 4.12, to follow) will be described first, followed by a special section on the latest developments in the measurement of strain and deflection. At this point, the reader should note that deflection, displacement or movement is measured in mm or microns (1 micron or μm = 1 × 10^{-6} m). Linear or direct strains are the ratio of the extension or compression movement in the direction of interest to the original length and are dimensionless ratios, usually measured in millistrain (1 mε = 1 × 10^{-3}) or microstrain (1$\mu\varepsilon$ = 1 × 10^{-6}.) Shear strains are angles of rotation measured in radians (where 360° of arc $= 2\pi$ radians. In structural measurements, shear strains are also measured in units of 1×10^{-3} or 1×10^{-6} radian.)

4.8.1 Determining principal and shear strains

In most cases, one is interested in the direct strain that arises in the direction of the measurement, e.g. along the axis of the member under investigation. However, sometimes the principal and shear strains are required, but the principal directions are not known. Here, a strain gauge rosette can be used, with strains measured simultaneously in three directions relative to a known direction, usually the axis of the member. The angles chosen can be completely arbitrary, but usually, for convenience and if space is available, directions of 0°, 45° and 90° or 0°, 60° and 120° are used, yielding strains ε_0, ε_{45} and ε_{90} or ε_0, ε_{60} and ε_{120}. Plate 4.2 shows a 0°, 90° electric resistance strain gauge rosette in which two strain gauges are overlapped one above the other to give as close as possible to two measurements, at a single point. The gauges shown in Plate 4.2 have a gauge length of 4 mm. This arrangement would be used when the directions of the principal strains are known. If the directions are unknown, they can be calculated from the measurements of a three gauge rosette. For a 0°, 45°, 90° rosette, (see Figure 4.5) by:

$$\varepsilon_1 = \tfrac{1}{2}(\varepsilon_0 + \varepsilon_{90}) + \sqrt{[\tfrac{1}{2}(\varepsilon_0 - \varepsilon_{45})^2 + \tfrac{1}{2}(\varepsilon_{45} - \varepsilon_{90})^2]} \tag{4.8a}$$

$$\varepsilon_3 = \tfrac{1}{2}(\varepsilon_0 + \varepsilon_{90}) - \sqrt{[\tfrac{1}{2}(\varepsilon_0 - \varepsilon_{45})^2 + \tfrac{1}{2}(\varepsilon_{45} - \varepsilon_{90})^2]} \tag{4.8b}$$

and the direction relative to that of ε_1 by

$$\tan 2\theta = [2\varepsilon_{45} - (\varepsilon_0 + \varepsilon_{90})]/(\varepsilon_0 - \varepsilon_{90}) \tag{4.8c}$$

The maximum shear strain is then

$$\gamma(\text{max}) = (\varepsilon_1 - \varepsilon_3) \tag{4.8d}$$

For a 0°, 60°, 120° rosette,

$$\varepsilon_1 = (\varepsilon_0 + \varepsilon_{60} + \varepsilon_{120})/3 + \sqrt{\{[\varepsilon_0 - (\varepsilon_0 + \varepsilon_{60} + \varepsilon_{120})/3]^2 + [(\varepsilon_{120} - \varepsilon_{60})^2/3]^2\}} \tag{4.9a}$$

and similarly (with a −sign before the √) for ε_3, and

$$\tan 2\theta = [(\varepsilon_{120} - \varepsilon_{60})/\sqrt{3}]/[\varepsilon_0 - (\varepsilon_0 + \varepsilon_{60} + \varepsilon_{120})/3] \tag{4.9b}$$

Having determined the principal strains ε_1 and ε_3, the principal stresses σ_1 and σ_3 are given by:

$$\sigma_1 = E(\varepsilon_1 + v\varepsilon_3)/(1 - v^2) \tag{4.10a}$$

$$\sigma_3 = E(\varepsilon_3 + v\varepsilon_1)/(1 - v^2) \tag{4.10b}$$

Where E is the elastic modulus of the material (in this context, of concrete or steel) and v is Poisson's ratio, usually taken as $\tfrac{1}{3}$ for steel and 1/6 for concrete.

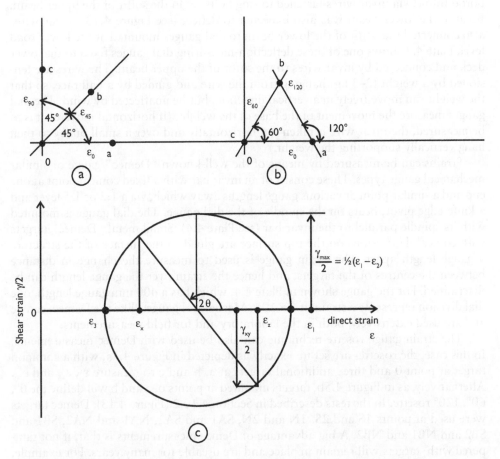

Figure 4.5 a and b: 0,45°, 90° and 0,60°, 120° Strain gauge rosettes, c: Möhr strain circle.

4.8.2 Mechanical methods for measuring deflection and strain

Methods most commonly used in the past (and still presently) include mechanical methods using mechanical dial gauges accurate to 0.001 mm either on their own or in combination with lever systems to give 2 to 5 times magnification of movement over gauge lengths of 50 to 500 mm. Dial gauges are used in both laboratory and field applications.

Deflection is usually measured directly by a dial gauge mounted on a stable base and with its spindle bearing on the underside of the deflecting specimen or member (e.g. the soffit of a beam). Alternatively, if no adjacent stable point is available to support the dial gauge, the deflection of a point can be transferred via a tensioned invar wire to a dial gauge mounted some distance away. For example in Figure 4.10, the deflections D1 to D8 were measured by dial gauges mounted at ground level, via invar wires fastened to the underside of the cantilever beam. In Figure 4.13, deflections P1, P2, P3 were measured by dial gauges on the deck of the lower beam of the portal frame, via invar wires fastened to ring bolts set in the soffit of the upper beam. Because the lower beam was also expected to deflect (see Figure 4.13), invar wires also connected the soffit of the lower beam to dial gauges mounted at the lower road level. Plate 4.3 shows one of these deflection-measuring dial gauges fixed to the lower deck and connected by invar wires to the soffit of the upper beam. The wires are tensioned by a weight (2–3 kg) hanging from the wire and guided by a ball race, so that the weight can move freely in a vertical direction, but be unaffected by wind. The dial gauges measure the movement of the base of the weight. (If horizontal movement is to be measured, the invar wire is taken out horizontally and over a small pulley to then hang vertically supporting the weight.)

Strains can be measured by means of the well-known "Demec" gauge or similar mechanical gauge types. These consist of an invar bar with a fixed conical point at one end and a similar point at various gauge lengths away which, via a 1:2 or 1:5 lever and a knife-edge pivot, bears on the spindle of the dial gauge. The dial gauge is mounted with its spindle parallel to the invar bar (see Plate 4.4). Small metal "Demec" targets with conical depressions on the top surface are glued to the surface of the structure a gauge length apart. The strain gauge is used to measure the change in distance between the centres of the targets, and hence the strain over the gauge length can be determined. For the gauge shown in Plate 4.4, which has a 400 mm gauge length, one dial division represents a strain of 5×10^{-6}. Mechanical gauges, like the Demec, can be and are used extensively, both in the laboratory and for field measurements.

The strain gauge rosette technique can also be used with Demec measurements. In this case, the rosettes are set up exactly as depicted in Figure 4.5a, with a common target at point 0 and three additional targets at a, b and c to measure ε_o, ε_{45} and ε_{90}. Alternatively, as in Figure 4.5b, targets mounted at points o, a and b will define the 0°, 60°, 120° rosette. In the tests described in Section 4.10.2 (Figure 4.13), Demec targets were used at points 1S and 2S, 1N and 2N, SA1 and SA2, NA1 and NA2, SB1 and SB2 and NB1 and NB2. A big advantage of Demec measurements is that, if not tampered with, targets will remain in place and are useable for many years. For example, in the tests described in Section 4.10.2, the Demec targets glued in place for the tests done in 1982 were still in place, and were also used in the 1988 tests.

4.8.3 Electrical methods for measuring deflection and strain

Three commonly used methods are LVDT's (linearly variable differential transformers), vibrating wire gauges and electric resistance strain (ERS) gauges (e.g., Hendry, 1964). These gauges are widely used both in laboratory testing and for field measurements.

The LVDT can be used in much the same way as a mechanical dial gauge. An iron core moves on a spindle inside a primary and two secondary coils, providing a variable output voltage (see Figure 4.6a). The movement of the core can be calibrated against the output voltage to measure the movement. Accuracies of 0.001 mm are possible. The LVDT can simply replace the dial gauge in measuring deflection or can be used as the movement sensor in a gauge, measuring movements between two points on the surface of a structural member. The advantages of the LVDT are that it can be read out, logged or recorded remotely whereas devices using dial gauges need

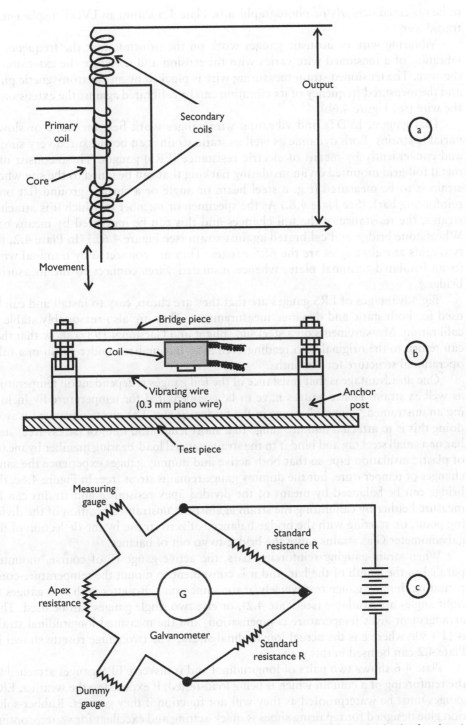

Figure 4.6 (a) LVDT displacement gauge, (b) vibrating wire gauge, and (c) balancing bridge for electric resistance strain (ERS) gauge.

to be observed ocularly or photographically. Plate 4.5 shows an LVDT displacement transducer.

Vibrating wire or acoustic gauges work on the principle that the frequency of vibration of a tensioned wire varies with the tension and therefore the extension of the wire. The tensioned strain-measuring wire is plucked by an electromagnetic pulse and the measured frequency of its vibration can be calibrated against the extension of the wire (see Figure 4.6b).

Dial gauges, LVDTs and vibrating wire gauges work best for static or slowly variable strains. Both dynamic as well as static strains can be measured very simply and conveniently by means of electric resistance (ERS) gauges. These consist of a metal foil grid mounted on an insulating backing that can be glued to the site where strain is to be measured (e.g. a steel beam or angle or a surface ground flat on a reinforcing bar). (See Plate 4.6.) As the specimen or member to which it is attached strains, the resistance of the foil changes and this can be measured by means of a Wheatstone bridge and calibrated against strain (see Figure 4.6c.) (In Plate 4.2, the two grids at right angles are the ERS gauges. They are connected by terminal wires to an insulated terminal plate, whence insulated wires connect to the measuring bridge.)

Big advantages of ERS gauges are that they are cheap, easy to install and can be used for both static and dynamic measurements. They are also reasonably stable in calibration. Measurements on a steel silo (Blight and Garstang, 1987) show that they can return to the original zero reading even after having been under strain in a fully operational structure for 9 months.

One disadvantage is that resistance of the foil gauges is dependent on temperature as well as strain. Hence gauges have to be compensated for temperature by including an unstrained "dummy" gauge in the bridge circuit. The most convenient way of doing this is to attach a dummy gauge to a short length (30 mm) of (stress-free) steel bar or a small steel tag and bind it to the strain-gauged load-bearing member by means of plastic insulation tape so that both active and dummy gauges experience the same changes of temperature, but the dummy gauge remains stress free. In Figure 4.6c, the bridge can be balanced by means of the divided apex resistance and strains can be measured either by calibrating the strain against the unstrained position of the dividing point, or, starting with the bridge balanced at zero strain, by the deflection of the galvanometer G as strains cause the bridge to go out of balance.

When strain-gauging reinforcing bars, the active gauge is, of course, mounted parallel to the length of the bar, and it is convenient to mount the temperature-compensating dummy gauge transversely at the same point. Rosettes with two gauges at right angles are available (see Plate 4.2), or else two single gauges can be used. This arrangement gives temperature compensation, and the measured longitudinal strain is $(1 + v)\varepsilon$ where ε is the actual longitudinal strain. The two gauge rosette shown in Plate 4.2 can be used in this way.

Plate 4.6 shows two pairs of longitudinal and transverse ERS gauges attached to the reinforcing of a column which is being load-tested. If exposed to the weather, ERS gauges must be waterproofed as they will not function if they get wet. Rubber-solution glue designed for repairing shoes is quick-setting and excellent for waterproofing strain gauges. Alternatively, aluminium foil-backed bitumen sealing strip can be used, and is very effective and long lasting. (See Plates 5.12 and 5.13, taken 20 years apart.

Strips of intact foil-backed sealing strip visible in Plate 5.12 are still intact where visible in Plate 5.13.)

Leads from each strain gauge pair can be taken to a convenient measuring or recording point where the strains can be read by hand or automatically logged or recorded. Leads should be protected from accidental damage and the weather by taking them through plastic electrical conduit tubing, from the position of the strain gauges to the measuring point. As leads have their own resistance, the lengths of the leads from the measuring and dummy gauges must be made equal. In Figure 4.13, E1 and E6 were either single gauges or gauges on two adjacent bars (e.g. E1 A and E1B).

4.8.4 Measuring temperature

Thermocouples provide a simple, inexpensive and robust method of measuring temperatures both in laboratory specimens and in structures under test. As diurnal thermal strains may be of the same order of magnitude as load-associated strains, it is essential to measure the internal and surface temperatures of any structure that is being load tested. The authors have used copper-constantan thermocouples for this purpose with great success and very few failures.

Reasonably stout thermocouple wire of 0.3–0.5 mm diameter should be used. Insulation is usually of braided glass fibre. The most successful and simplest temperature-sensing junction is made simply by stripping the insulation from the wires over a length of about 10 mm and twisting the wires tightly together using pliers. Junctions of this form have never failed the authors. For steel structures, thermocouples can be installed simply by sticking the junction against the steel using aluminium-backed, bitumen sealing strip. Alternatively, a metal or plastic saddle held in place with small self-tapping screws can be used to clamp the junction against the steel. As steel is a good heat conductor, there is no concern about lateral or transverse temperature gradients local to the junction. The junction must be covered, however, or it may respond to rapid surface temperature fluctuations caused by the wind. The aluminium backing or whatever is used to cover the junction should preferably be painted or coloured the same colour as the surrounding steel to avoid differences in albedo, i.e. heat reflectivity.

For concrete structures there is often a need to know temperature gradients through the concrete thickness. Hence thermocouples have to be installed at the outer surface and at various depths within the concrete thickness. The simplest way of doing this is to drill a hole with a masonry drill (about 10 mm diameter) into the wall. The thermocouple junctions are then taped in their correct relative positions along a thin wooden dowel stick. The hole is filled by injecting a liquid sand-cement grout or alternatively a sand-epoxy resin grout and the dowel and the thermocouples it carries are pushed into the hole, displacing the surplus grout and sealing the thermocouples into the concrete at the required depths. As the wooden dowel is itself a good insulator, it may be left in place in the hole. If a single thermocouple is installed on the outside surface of a reinforced concrete structure, it is best to protect the junction from the wind by placing it in a shallow (5 mm deep 3 mm diameter hole drilled in the wall. The hole is plugged with sand-epoxy mix or epoxy putty and the thermocouple wire is clamped in position by means of a saddle and steel nails.

In Figure 4.13, T1 to T6 were sets of thermocouples set at various depths (5, 250 and 500 mm) from the surface of the concrete to provide measurements of temperatures at depth and temperature gradients in the concrete. Plate 4.7 shows the front end of a dowel and thermocouple probe that has been in soil for a year and which still functions perfectly. The twisted junctions of two of the thermocouples have been exposed for the photograph and are seen near the prongs of the two two-pin plugs used to connect the two thermocouples to the measuring bridge.

The thermocouple wires are taken down the outside of the test structure to a convenient reading position. It is important to protect the wires from abrasion otherwise, if the insulation gets abraded away for example, as the wire is blown back and forth by the wind, and exposed copper and constantan wires touch at a point between the sensing junction and the measuring point, the recorded temperature may represent that at the point of touch, rather than that at the junction.

The thermocouples are read by means of a thermocouple bridge or an automatic data logger, or can be recorded continuously by means of a chart recorder. Thermocouples can easily be calibrated over the expected range of measurement by direct comparison with an accurate mercury-in-glass thermometer.

4.8.5 Measuring rotation or change of slope

Rotation of a joint or the slope of a member can be measured by means of a sensitive spirit level tube mounted on a strip of invar that is pivoted on a knife-edge at one end and attached to a screw micrometer at the other. It is possible to obtain level tubes that can be leveled to an accuracy of 50×10^{-6} radian. The bubble is mounted on a bracket bolted to the surface of the member where it is intended to measure rotation. The bubble is centred in the tube and then re-centred once the rotation has occurred. The distance measured on the micrometer, at the end of the tube mounting, necessary to re-centre the bubble, divided by the distance between the axis of the micrometer and the knife-edge, gives the joint rotation.

Plate 4.8 shows a spirit level or bubble gauge mounted on a bracket that is bolted to the side of a structure. In this case, because of lighting difficulties, the micrometer screw to the left has been augmented by a dial gauge to give two independent measurements of tilt or rotation (Two thermocouple wires cross the field of view below the spirit level gauge.) These gauges can also be used in the laboratory.

In Figure 4.13, spirit level gauges were mounted on either side of knee J at points B1 to B4 to measure in-plane rotation. If rotation at B2 and B4 had differed from that at B1 and B3, it would have shown that moment continuity at knee J had been lost, and vice versa. The level gauges at B5 and B6 measured out of plane rotation as the load on either side of the portal varied.

4.8.6 Recent developments for in situ measurement of displacement, rotation and strain in structures

(An overview by Alex Elvin, Professor of Civil Engineering, University of the Witwatersrand.)

The deflection of structures subject to externally applied loads, thermal variations and internal processes have been measured by a variety of approaches and multiple commercially available instruments. In this section, only the more commonly used, tried and trusted sensors and instrumentation systems are discussed. Over the past 50 years either mechanical or electrical systems have been employed. For example, deflections have been measured by mechanical dial gauges based on spring and gear mechanisms, while to determine slope or small rotations bubble levels have been utilized. The Linear Variable Differential Transformer (LVDT) and the Rotary Variable Differential Transformer (RVDT) are standard electrical instruments to measure linear and rotational displacements. Both rely on electromechanical transduction and put out an alternating current that is linearly proportional to the magnitude of displacement or rotation. Electrical systems are less subject to user reading error (*e.g.* parallax), and are thus more repeatable.

Strain deformations have also been measured using mechanical and electrical devices. By mounting targets on the structure and using a Demec (demountable mechanical) gauge (Plate 4.4), strains can be calculated. This device uses the dial gauge to measure deflection over a fixed and known length. Since the Demec targets are placed at least several centimetres from each other, the resulting reading is an average strain over this distance. The longer the instrument the more accurate it becomes, at the cost of averaging the strains occurring in the structure between the targets. Strain gauges are much smaller, and give a more accurate point wise reading. Three main varieties of strain gauges exist: electrically resistive, piezo-resistive and piezoelectric. By attaching strain gauges (on the surface or within the member), the deformation of the structure results in a change in the electric property of the gauges. It is this property change that is sensed.

In general, mechanical gauges have the following limitations. Only surface displacements or strains can be measured. There must be access to the structure for placement and reading of the instruments. These limitations can be severe if the structural component under investigation is inaccessible. Only slow varying or static readings can be taken. The benefit of such systems is the low cost of the data obtained.

Some electrical sensors on the other hand, can be embedded into the structure. After placement, and provided the connecting wires do not have an effect on the measured signal (e.g. noise), the electrical output has to be processed (usually by a filtering and buffering electrical circuit). The measurement can then be displayed in an analogue format, or more recently the data is converted from analogue to digital representation and then logged. This makes instrumenting the structure costly, thus restricting the number of sensor nodes. With the advent of multiple channel analogue to digital converters, the dynamic response of structures could be measured at multiple points.

Modern electrical sensor transducers can be attached to radio links which can transmit the data wirelessly. Without wiring, deployment of these sensor nodes has become easy and economic. The sensor nodes can form a network and whole structures can now be monitored. There are three main limitations to RF linked sensors: (a) if the radio link is disrupted, important data can be lost – this is especially true for relatively rare events such as earthquakes; (b) the rate at which data is acquired is limited by the bandwidth of the radio link – this is only important for very fast events such as impacts; and (c) powering the sensor nodes.

Sensor nodes can be powered using batteries or mains supply. The former raises the question of battery longevity, while the latter requires reliability of the power supply and wires. At least two wireless powering schemes have been used: (a) using photovoltaic cells, and (b) inductive powering. A photovoltaic cell can be attached to the remote sensor node(s) and can recharge the batteries. The main disadvantages of this system are that the cells require light, have to be positioned perpendicular to the light source for optimal operation and have to be bigger than several square centimetres (approximate peak photovoltaic power density is 15 mW/cm^2). This prevents the sensor node from being embedded in the structure, and the system requires special handling since existing photovoltaic cells are brittle and susceptible to performance degradation with dirt or dust collecting on them. The second method of powering relies on an inductive source passing close to the sensor and energizing the node. The sensor node containing a receiving loop antenna can be embedded within the structure and be small in volume. The negative aspect of this powering scheme is that the inductive source has to be relatively close to the sensor to be effective. Unless this source is very slow or stationary the sensor nodes operate in "burst" mode and batteries cannot be recharged. In this scheme, the inductive antenna can receive power from the travelling source, the sensor can take the measurement, and the results can be transmitted to the same source.

To solve the powering issue, several research groups are engaged in developing small volume self powered sensor nodes. In most research devices, active (also known as "smart") materials are used to harvest ambient energy, *e.g.* from the vibration of the structure during exploitation, to either recharge the batteries or power the sensor nodes directly (e.g. Elvin et al., 2006). If successful, the sensor nodes would indeed become fully wireless. To date, the amount of harvestable energy is too small to drive the wire-less electronic sensors.

In the last decade structures have been instrumented with commercially available accelerometer and gyroscopic sensor nodes. These accelerometers have become low cost, small (4 × 4 mm is typical) and can measure in three directions. The sensors are based on a new solid state technology called MEMS (or micro-electro-mechanical-systems) and are manufactured in a similar way to electronic microchips. MEMS accelerometers can have a variety of ranges from 2 g to determine tilt (with respect to gravity) all the way to 250 g to measure impact events. Accelerometers have been used on structures in earthquake zones, as well as structures subject to dynamic loading (e.g. Elvin and Uzoegbo, 2010). Although accelerations can be integrated twice to derive deflections, due to sensor noise, these deflections are not accurate. A variety of research schemes exist to improve the prediction of structural deflection based on measured accelerations. Accelerometers are particularly useful to capture the dynamic behaviour of a structure. From the dynamic response, the potential for, or extent of structural damage can be determined.

Also in the last decade and a half, low cost optic or light fibres have been used to measure deformation in structures (e.g. Udd, 1996). A small diameter fibre, (the core can be substantially less than 1 mm in diameter), identical to the ones used in the telecommunication industry, is embedded into, or attached to a structure. Fibres can be of any length. A laser is shone into a stranded fibre, and depending on the deformation, the transmitted or reflected light intensity is affected. Relatively sophisticated electronic circuits measure and log the extent of this effect and hence the deformation. Since the light fibre is embedded along its length, an average deformation reading is obtained. Accuracy can be up to ±0.02 mm for gauge lengths of up to 30 m. This is both an advantage and a disadvantage. The overall health of the structural member can be ascertained using this distributed sensor. However if there is damage, or the member is not behaving to specification, the sensor cannot determine where the problem is occurring. To overcome the distributed sensor limitation, the light fibre can be processed by etching Bragg gratings along its length. Each grating forms a sensor node and can be interrogated individually to give locations of the occurrence of strain to within ±0.15 mm. Unfortunately, etching these gratings is extremely expensive. The light fibres in general are brittle and require careful handling and placing. The main advantage is that light fibre based sensors can be extremely sensitive and versatile. To emphasize their versatility, light fibres can be processed and hence used to detect the presence of specific chemicals in low concentrations.

Recently, Ground Positioning Systems (GPS) have been used to monitor structures such as bridges, tall buildings and offshore platforms (e.g. Meo and Zumpano, 2005). Although the GPS hardware technology itself is incapable of producing measurements that are sufficiently accurate for displacement measurement, advances in data-processing software has improved accuracy to the order of a few millimetres. With such accuracy, only large structures which experience large deflections compared to this accuracy can be instrumented with GPS nodes. Further, position can be measured in three-dimensions at a slow rate of 10 Hz. Hence, at the present time, GPS is limited to large structures with low natural frequencies. These sensor nodes can only be attached to the top surface of the structure which is open to the sky. The advantage of this system is that it is straight-forward to use, and is wireless.

Many non-contact sensing systems are currently being researched. One such system relies on applying the principles of photogrammetry in digital photography. By placing markers on a structure, using high resolution digital cameras with lenses, and software that automatically identifies the markers, first displacements and then deformations are computed. It is still unclear whether sufficient sensitivity and hence accuracy can be achieved. Of course only surface deformations, and visible points can be studied. The main advantage is the low cost remote markers that can be monitored when the structure is loaded statically or dynamically.

It must be emphasized that many sensors and systems based on a variety of physical principles and properties exist. Only some of these have been briefly introduced. New instrumentation schemes are actively being researched, not only to improve accuracy, but also to overcome the limitations of various systems.

4.8.7 Testing by ultra-sonic pulse velocity (UPV)

One of the most frequently used non-destructive exploration techniques for assessing the extent and severity of AAR damage to concrete structures is that of measuring the ultra-sonic pulse wave velocity or UPV. This has been used both in the laboratory to track the course of deterioration as a result of AAR (e.g. Akashi, et al., 1986, Amasaki and Takagi, 1989, Gallias, 2000) and in the field (e.g. Kojima, et al., 2000, Ono and Taguchi, 2000). Section 3.5 has illustrated that UPV measurements give an excellent indication of the state of integrity or disruption of concrete, but are very poor at indicating concrete strength.

The velocity of a sound wave is a function of the dynamic elastic modulus E_d and density or unit mass ρ of the material transmitting the wave. The wave velocity V is given by

$$V = \sqrt{(KE_d/\rho)} \tag{4.11}$$

where K is a function of Poisson's ratio v: $K = (1 - v)/\{(1 + v)(1 - 2v)\}$.

For an incompressible material, $v = 0.5$ and $K = \infty$, for concrete $v \approx 0.2$, $K = 1.11$ and is dimensionless.

$$[\text{Dimensionally } [V] = \sqrt{[Nm^{-2}]/[kgm^{-3}]} = [kgms^{-2} \cdot m^{-2} \cdot kg^{-1}m^3]^{\frac{1}{2}} = [ms^{-1}]$$

An ultra-sonic pulse velocity tester has a pair of terminals connected by flexible cables to a wave pulse generator, a pulse receiver and a timer. The pulse generator transmits a series of ultra-sonic wave pulses from the transmitter terminal, while the receiver, at the other terminal, picks up the arrival of the pulse after being transmitted through the concrete, and the timer measures and displays and/or records the transit time of the pulse from transmitter to receiver.

Both transmitter and receiver terminals must be well sonically connected to the concrete surfaces. On smooth concrete surfaces the terminals can be bedded in by means of a thin layer of petroleum jelly or a heavy grease (just as with a medical ultra-sound apparatus). Irregular surfaces may require rubbing down to smoothen them, or skimming with a layer of plaster of Paris before applying the petroleum jelly to achieve good sonic contact.

Plate 4.9 shows a typical UPV apparatus. In this case, the specimen being measured is a core of AAR-affected concrete that has been smooth sawn in half longitudinally. The surface contains no visible cracks The short cylindrical transmitter and receiver terminals have been bedded in with petroleum jelly that can be seen as a slightly darker "halo" around each terminal. The terminals are connected by means of cables to the housing containing the pulse generator, the pulse receiver and the timer. The timer read-out displays the transmission time, in this case 41.1 microseconds (41.1×10^{-6} s). As the transmission distance between the two terminals is 150 mm, the ultrasonic pulse velocity can be calculated as

$$V = 0.150/(41.1 \times 10^{-6}) = 3650 \text{ m/s or } 3.65 \text{ km/s}.$$

This is not a particularly high UPV, but the core has been damaged by AAR and is completely dry, hence any cracks or voids crossing the transmission length would be air-filled and would retard the ultra-sound wave. Referring to equation (4.11), it will be seen that the transmission velocity depends on the ratio E_d/ρ. Air has a very small density, and a small modulus E_d, in comparison with water, which has much larger ρ and E_d values. However, the ratio of E_d/ρ is much larger in the case of water, and hence a water-filled crack will transmit a sonic pulse better than an air-filled one. For a similar reason, a steel reinforcing bar traversed by an ultrasonic pulse will accelerate the transmission of the pulse.

It should also be noted that if the wave velocity has been measured, the dynamic elastic modulus can be found from

$$E_d = \rho V^2/K \approx 0.90\rho V^2 \text{ for concrete} \tag{4.11a}$$

For example, if $V = 4000$ m/s and $\rho = 2200$ kg/m³

$$E_d = 0.90 \times 2200(4000)^2 \text{ kgm}^{-3} \cdot \text{m}^2/\text{s}^2 = 31.7 \times 10^9 \text{ N/m}^2$$

$E_d = 31.7$ GPa. (Note that dynamic moduli are usually larger than moduli measured in static tests.)

Depending on available access, the purpose of the measurements and the available length of the cables connecting the terminals to the transmitter and receiver housing, various patterns of transmission can be used, as shown in Figure 4.7. Plate 4.9 shows (a): indirect, same face transmission being used. The simplest method is (b): direct transverse transmission, but restrictions on accessibility may require (c): across corner transmission to be used.

In every case, and regardless of the arrangement of terminals, the transmission path length (between the centres of the terminal faces) must be determined as accurately as possible. This is easy on laboratory specimens, but may be difficult on structures in the field. Wherever possible, distances must be taped directly to the nearest mm. Dimensions taken from construction drawings cannot be relied on for accuracy.

In all UPV measurements, the reinforcing should at first be located by means of a covermeter or bar detector, and transmission paths should not be crossed by rein-

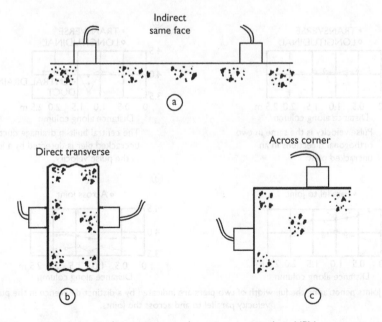

Figure 4.7 Various arrangements of transmitter and receiver terminals in UPV measurements.

forcing, or run parallel to adjacent reinforcing. Thus, it is unwise to use the indirect method along the length of a beam or parallel to the height of a column, unless the design drawings and the bar detector show that the transmission path is clear of reinforcing.

Figure 4.8 shows a set of measurements made by direct transmission on four rectangular columns, (a) and (c) measuring 1.5×2.25 m and (b) and (d) 2.0×2.25 m. In every case, the outside measurements were taken 0.25 m from the corners of the columns, to avoid longitudinal reinforcing and positions were also chosen, using a covermeter, to avoid lateral steel. In (a) the column was apparently completely uncracked, and the UPV measurements show that this was the case. In (b) the column was also uncracked, but was known to contain a 200 mm diameter rigid PVC drainage pipe. This is clearly shown up by the UPV measurements. In (c) and (d) the columns were traversed by a halving construction joint parallel to the long side. Transmission parallel to the joint was uninterrupted, but the interruption of the construction joint shows up clearly in the other direction.

Figure 4.9a shows the results of two UPV surveys of the beam of a reinforced concrete portal (that is shown in Figure 4.13). The measurements were by direct transverse transmission with one instrument (using two hydraulic lifts for access to the measuring points, one on each side of the 2 m deep beam), at exactly the same points. (In 1988, these were still marked by the grease used to bed in the terminals 5½ years previously. As the diagrams show, the results of the two sets of measurements were completely different, with the majority of measurements well below 4 km/s in 1982 and the reverse being the case in 1988. The only possible explanation can be that in October 1982, at the end of the dry season, the concrete was considerably drier

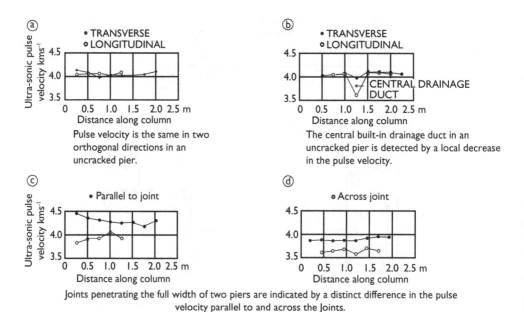

Figure 4.8 Demonstration of capability of UPV tests to detect uniform sound or cracked concrete.

than it was in April 1988, the end of the wet season. The concrete forming the head of the column appears to have deteriorated between 1982 and 1988, and this is also possible.

Whereas Figure 4.9a shows detailed UPV measurements made 5½ years apart, Figure 4.9b (Cole and Horswill, 1988) shows more comprehensive, but less detailed UPV measurements on the downstream surface of the AAR-damaged mass concrete gravity Val de la Mare dam in Jersey. These five surveys were made over a period of 11 years. The authors concluded that "although there are considerable variations in the average sonic velocity between surveys, the variations are not systematic with time..." "... there is no evidence to suggest an underlying trend of significant deterioration of the concrete with time."

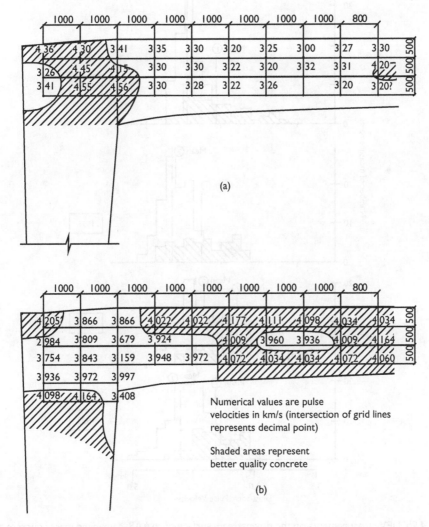

Figure 4.9a Ultrasonic pulse velocity survey of upper transverse beam of portal: (a) October 1982 survey; (b) April 1988 survey. Dimensions are mm.

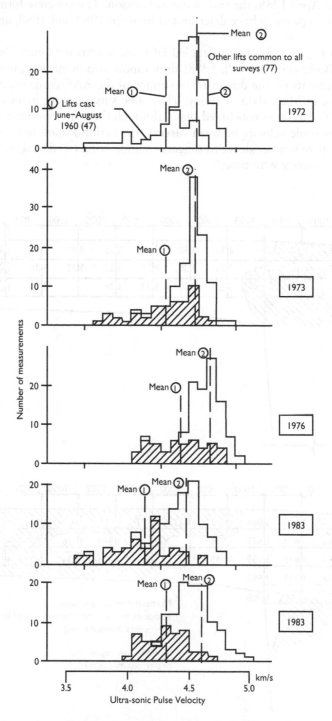

Figure 4.9b UPV measurements on the downstream surface of an AAR-damanged mass concrete gravity dam from 1972 to 1983 (Cole and Horswill, 1988).

The overall conclusion regarding UPV measurements is that they are useful in locating invisible disruption of concrete, built-in drainage ducts, etc., but give no idea of the concrete's strength. UPV measurements cannot be used successfully to track progressive deterioration by AAR, as they are influenced too much by factors such as the water content of the concrete, the progressive filling of cracks with AAR gel, etc.

In their 1992 technical guidance on the structural affects of AAR, the British Institution of Structural Engineers summed up the capability of UPV measurements to detect damage caused by AAR to concrete as follows:

"The measurement of UPV is sometimes a useful non-destructive testing method. Changes in pulse velocity ... can indicate different degrees of deterioration .... However, the pulse can be short-circuited by reinforcement so that the procedure is not readily applicable to reinforced elements ... While pulse velocity falls rapidly in the early stages of AAR expansion it changes little as further expansion develops.

The UPV may be markedly reduced by micro-cracking ... However, where cracks are filled with gel and/or water, the UPV may not be reduced ... after an initial decrease of UPV due to AAR cracking, there can be a substantial recovery as cracks become filled with gel." (The present authors have replaced the acronym ASR with AAR.)

4.9   PLANNING, PREPARING AND PERFORMING AN IN SITU LOAD TEST ON A STRUCTURE

(The authors have planned and successfully performed more than 20 major field measurement projects between 1965 and 2010 and consider that they are qualified by experience to write on this subject.)

4.9.1   The history of the structure

Test loading a structure to assess its condition of structural safety is always a complex operation that requires meticulous planning. Usually, concern about the safety of the structure has grown over a period of years, possibly with the appearance of progressive cracking and, in this context, the diagnosis of AAR as the cause of the cracks. There will therefore be a history of deterioration, either documented by means of photographs and measurements of crack length and width, or at least by written inspection reports. These should all be studied in conjunction with available design drawings and calculations, if these still exist or can be found, as well as construction records, test cube or cylinder results, mix designs, the source of the aggregate and cement, etc.

4.9.2   Objectives, extent of testing and preliminary information-gathering

The objectives and extent of the testing must be discussed and agreed with the owners of the structure and a preliminary proposal and budget prepared, discussed, modified and agreed. Depending on what information about the structure is available, this preliminary stage may include taking test cores to help establish the strength and elastic modulus of the concrete, exposing reinforcing in selected areas to check on the type of steel, on numbers and presence of bars and cross-sectional areas of steel.

It may even be necessary to measure up the structure to determine spans, heights and overall dimensions of members. If design calculations cannot be located, or prove on examination to be inadequate, an analysis of the structure for the original, or currently required design loading must be undertaken. Predicted load versus strain and movement predictions must be prepared.

With the new information available, a revised structural analysis and a revised proposal for the test, with a revised budget, can be prepared, discussed and agreed with the owner.

4.9.3 Detailed planning – choice of date and time, lighting and access

In order to test load a structure, it will usually be necessary to put it out of use for at least 48 hours. This requires selecting a time for the test which falls in a two or three day quieter period, e.g. a long week-end (like Easter) when decommissioning and diverting activity and traffic will cause less disruption. As structures afflicted by AAR are invariably exposed to the weather, a period of moderate temperatures when the weather can be expected to be stable and clement should be chosen. This usually means a period near the spring or autumn equinox.

As mentioned earlier (Section 4.8.4), rapidly changing temperatures can produce strains and movements in a reinforced concrete structure that are comparable in magnitude to the load-associated effects. Hence before the test begins, a period of slowly cooling temperatures, leading to a period of stable temperature is required to allow temperatures within the structural members to equalize, stabilize and remain stable while the test is in progress. This means that the best time to start testing is 22 h to 23 h, with the test running until 03 h to 04 h the next morning. Easy, safe access must be provided so that all instrument sites can safely and easily be accessed. This, in turn may require that secure and safe scaffolding be erected to provide the access. Good lighting must also be arranged so that those observing the instruments can move around safely and clearly see what they are doing, observing and noting. It is essential to allow adequate time to set up the access system, lighting, auxiliary power, toilet, etc. as this invariably takes longer than predicted. Setting up the instrumentation cannot begin before access has been established. Ideally, the facility being tested should be closed off to public use at dawn on the test day to allow at least the afternoon for setting up the instrumentation in good daylight.

Where possible, a weather-proof, well-lit "command post" such as a caravan must be provided, to house as many of the logging and recording instruments as possible. If manual and ocular instruments such as dial and Demec gauges are used, there should be two dedicated observers, one to operate and read the gauge (which in the case of the Demec requires two steady hands and a pair of good eyes), the other to record the readings. Obviously, these instrument sites must be easily accessible and well-lit.

4.9.4 Loading system, stages of loading, predicted and actual movements and strains

If there were no chance that the structure might not pass the load test, the test would not be necessary. Hence loads must be applied in increments (usually 4 to 6) so that the loading can be stopped and removed, if necessary, at any stage. The resulting strains and movements must be checked against predicted values after each load increment. A complete set of measurements (including temperatures) must be taken and compared with predictions based on the structural analysis described at the end of Section 4.9.2. If all measurements are within or less than the theoretical predictions, a period (usually 30 to 45 minutes) should be allowed and the set of measurements be repeated, both as a check on the first set and to observe any progressive or creep-type movements. If the second set of measurements agrees with the first, the next load increment can be applied. Loading in this way is safe and each load increment takes about 1 hour, giving 4 to 6 hours for loading and 2 hours for unloading. If the test starts at 22 h, it should be complete by 06 h, or daybreak at the equinoxes. This will leave the early morning to dismantle the instruments, return the loading system to its source, dismantle and remove the scaffolding and return the structure to service by noon or early afternoon.

The required magnitude of the test loading may be prescribed by local structural codes of practice and should amount to at least the required design load (although, as described in Section 4.4.1, it may prove physically impossible to apply the design load). The British Standard BS 8110 (1995), for example, as a final total test load, requires the greater of:

design dead load + 1.25 design imposed load, or
1.125 (design dead load + design imposed load).

Methods of applying the test load will vary with the type of structure. For structures like ware houses, public concourses or parking garages, loading could be applied by means of concrete blocks, stacked bricks, water bags, etc, while for highway structures, both for static and dynamic loads, loaded and weighed trucks, either all of one type, or of different types to get a variation of axle spacings, could be used.

4.9.5   Briefing the testing team

The testing team is rather like an orchestra: each person or observing team has their own instrument to observe, it must be observed or activated at exactly the right times, and everyone must be alert and ready at the due time for their next action. There must therefore be a team leader (conductor) to co-ordinate the various actions, to signal when they are due to occur, to continually monitor progress and the outcomes of the various actions, and to modify the programme if anything untoward occurs. Each member of the team must have a copy of a detailed programme and must be completely aware of the part he or she must play. On a big site there must be a system of mobile distance-communication by means of two-way radios or cell phones. For example, the person marshalling the loaded trucks must be in continuous connection with the team leader, as well as the truck drivers.

The site of the test should be occupied as early as practicable on day 1 of the test period with erection of scaffolding starting as soon as possible and the command post also in position as early as possible. Mounting of instruments should also start, as soon as their planned locations become accessible. The lighting and power systems should also be installed and tested as early as possible. In remote locations, power for lighting and instruments may have to be supplied by a generator (with a standby generator) and even in urban locations it is wise to have a standby generator in case of a problem with the mains supply. Plate 4.10 shows the portal frame described in Section 4.4.1 being prepared for test loading in 1982.

Once preparations are complete and the team is waiting to start the test, an on site barbeque evening meal (without alcoholic drinks) is an excellent way to boost flagging spirits and improve team enthusiasm. A constant supply of tea and/or coffee throughout the night is essential, as is an onsite toilet.

4.10   "SPECIAL" OR "ONCE OR TWICE OFF" TEST LOADINGS OF COMPLETE STRUCTURES

4.10.1   Motorway double-cantilever structures: (northern cold-temperate coastal climate)

Imai et al. (1986) carried out comparative test loadings on four of the line of single columns supporting the double-cantilever overhead structure of the Hanshin express way in Kobe, Japan. Figure 4.10 shows the layout and principal dimensions of a typical column and cross-head. The structures were completed in 1975. In 1982 cracking caused by AAR had been observed in a number of the columns. Some of the cracks were up to 5 mm wide. The cracks were first noted in 1979, four years after completion of construction.

Two of the columns subjected to test had been affected by AAR and two had not. Photographs and diagrams show that the surfaces of the affected double-cantilever cross-heads were covered in a network of surface cracks and comparative measurements of ultrasonic pulse velocities on unaffected and affected piers had varied from a maximum of 4.20 km/s to a minimum of 1.92 km/s. Comparative measurements on cores cut from unaffected and affected cross-heads showed that core strengths had reduced from about 36 to 23 MPa, while elastic moduli had reduced from 28 GPa to only 6 GPa.

Test loadings of up to 80% of the design live load were carried out in 1984. To apply the most severe loading to the cross-head cantilevers, loaded trucks were placed in the outer lanes of the highway and the deflections of the cross-head measured. Both symmetric and asymmetric loading was applied. For symmetric loading, three loaded trucks were located on either side of the cross-head, as shown in Figure 4.10, giving a load of 680 kN on each side of the double-cantilever. For unsymmetric loading, three trucks were located on one side only. Figure 4.11 compares measured and calculated deflections for the most severely damaged pier (P-42) for both the symmetrical and unsymmetrical loading cases. It was deduced that the effective elastic modulus of the AAR-affected concrete was actually 90% of that of the sound concrete; and that the stiffness and load carrying capacity of the deteriorated portions of the structure were similar to those of the sound structures, at the time. This shows that the results of tests on cores taken from a damaged structure are not necessarily representative of the behaviour of the entire structure.

It would have been logical to repeat this test some years later to see if any change in structural integrity had occurred, and possibly such tests were made, but the results

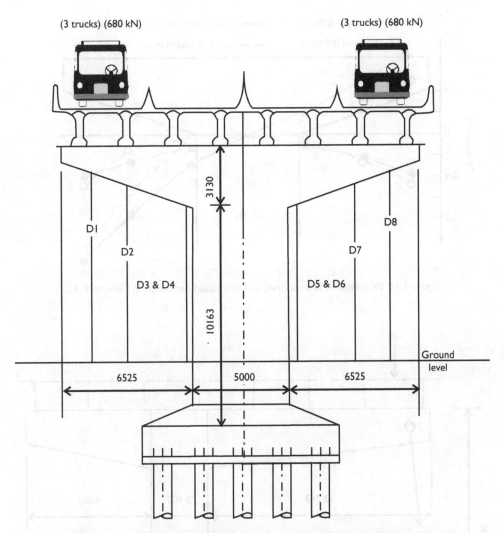

Figure 4.10 Schematic layout of loading test on supporting columns and double cantilever beams of Hanshin expressway in Japan. Dimensions are in mm and D1 to D8 indicate locations of deflection measurements.

do not appear to have been published. However, large lengths of these overhead structures were destroyed in the 1995 Kobe earthquake which toppled many of the supporting columns and cross-beams. Plate 4.11 shows a collapsed section of the Hanshin expressway being demolished after the earthquake. One of the double cantilever supporting columns can be seen in the centre of the photograph.

Figure 4.12 (Ono, 1989) shows the layout and dimensions of a second series of column and double cantilever beams supporting an elevated roadway. A number of these supporting structures, located in the south western part of Japan, near Osaka and Hiroshima, had deteriorated as a result of AAR. It was decided to investigate two

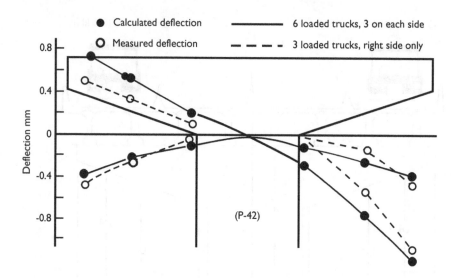

Figure 4.11 Distribution of measured and calculated deflection in Column P-42.

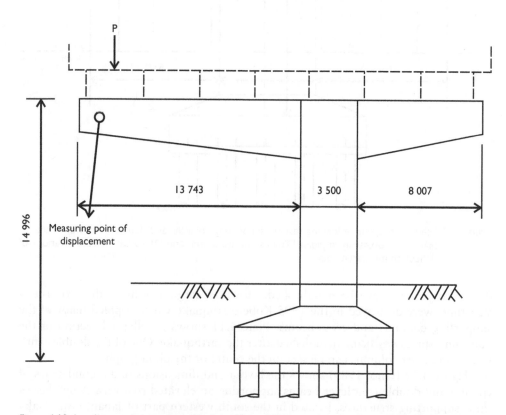

Figure 4.12 Loading test on asymmetrical double cantilever and supporting column. Note that both cantilever spans surpass those shown in Figure 4.10, with the left-hand span being more than double the span of the Hanshin structure. Dimensions are in mm.

structures by full scale test loading, one that had deteriorated and one that had not. The comparison was expected to tell the degree of deterioration suffered by the one that had suffered AAR attack.

In both cases, the longer cantilever was loaded by parking a weighed loaded truck as close as possible to the end of the cantilever, and measuring the static deflection, as shown in Figure 4.12. A dynamic load was then applied by driving a truck weighing 405 kN over the structure at 60 km/h. The static load amounted to 596 kN while the static equivalent of the dynamic load was 323 kN. The measured deflections are shown in Table 4.2, together with back-analysed values of the elastic modulus of the concrete and values measured on a core (or cores – it is difficult to tell from the paper) taken from the AAR-affected structure.

Table 4.2 shows (line 4) that the back calculated value of E for the AAR-damaged concrete was 87% of that of undamaged concrete, whereas the value measured on a core was an unrealistic 28% (line 5). This shows again that the performance of structures damaged by AAR is usually much better (lines 1 and 2) than would be expected from their appearance and from tests on limited numbers of cores taken from the structure.

It is also interesting to note that proportionately (line 3), a static load on either structure caused a deflection that was larger than the static equivalent of the dynamic load. This agrees with an earlier observation (Blight 1976) which showed that aerodynamic lift on the underside of a loaded truck moving at 60 km/h reduced the loads transferred by the wheels to the road surface by 20 to 35%. In this case the reduction at 60 km/h appears to have been 13%.

4.10.2 Motorway portal frame (southern warm-temperate, water deficient continental climate)

Blight (1983) and Blight, et al. (1989) carried out two full-scale loading tests, six years apart, on a reinforced concrete portal frame that was showing apparently severe deterioration as a result of AAR. The structure, one of a series supporting a double-decker stretch of urban freeway, had been designed in 1963 when the occurrence of

Table 4.2 Results of loading tests on highway support structure shown in Figure 4.12.

		Condition of structure	
		Unaffected	Affected by AAR
1	Static deflection of cantilever under load of 596 kN	2.40 mm	2.75 mm
2	Dynamic deflection under load of 323 kN moving at 60 km/h	1.47 mm	1.69 mm
3	Dynamic deflection × 596/323	2.71 mm (113% of 2.4 mm)	3.12 mm (113% of 2.75 mm)
4	Back-calculated value of E	40.5 GPa (100%)	35.3 GPa (87%)
5	E measured on AAR-affected core		11.2 GPa(28%)
6	Static deflection × 40.5/35.3	2.75	Checks line 4
7	Dynamic deflection × 40.5/35.3	1.69	Checks line 4

AAR was unknown in South Africa. The layout of the frame and the locations of the measurements are shown in Figure 4.13. The comments in Section 4.4.1 refer to this portal frame. Plate 4.10 is a view of the portal being prepared for the 1982 test and Plate 4.12 shows the loaded trucks, providing the staged loading, being arranged on the upper deck. Plate 4.6 shows the surface appearance of the concrete forming the upper knee J referred to in Figures 4.13 and 4.14. The hole in the vertical face was used to access the vertical tension reinforcing in the knee for the purpose of strain gauging the tension reinforcing in it (see Section 4.8.3). Prior to the loading tests, an elastic finite element analysis of the frame had been made using a reduced value of elastic modulus for the concrete, that had been established by means of laboratory measurements on cores taken from the structure. The deformed shape of the portal under load resulting from the analysis is shown in Figure 4.14. Measured deflections, rotations and strains were then compared with the previously predicted quantities. In every case, close agreement was found between prediction and measurement. Over the short time duration of each of the loading tests, the structure behaved almost completely elastically and deformations were nearly fully recovered on removal of the load. Figure 4.15a shows the predicted and measured load-deflection curves for midspan of the upper beam of the asymmetrical portal and Figure 4.15b shows the corresponding predicted and measured strains in the midspan tensile reinforcing. The agreement obtained between strains, rotations and displacements predicted by analysis on one hand and observation on the other was excellent, and the results of the two tests, made six years apart in 1982 and 1988 are almost indistinguishable. (In Figure 4.15, experimental points are shown for the 1988 test. The 1982 test results are

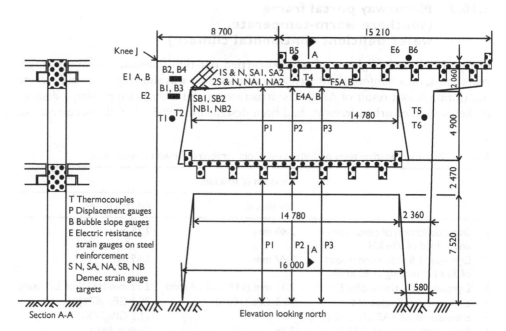

Figure 4.13 Elevations of portal frame showing positions and types of instruments used in full-scale load tests made in 1982 and 1988. (Dimensions are in mm.).

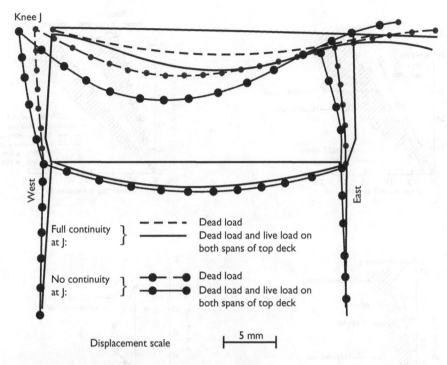

Knee J

West

East

Full continuity at J: }
- - - - Dead load
———— Dead load and live load on both spans of top deck

No continuity at J: }
—•—• Dead load
—•—• Dead load and live load on both spans of top deck

Displacement scale   ⊢ 5 mm ⊣

Figure 4.14 Calculated deformed shape of portal under load.

shown as chain-dotted lines. The location of the points of measurement, P2, E4, E5, SA, NA, 2 N are shown in Figure 4.13.) Note from Figure 4.15a that even though the knee J of the portal was apparently badly damaged by AAR, it still behaved as if fully continuous. Note from Figure 4.15b that the beam of the portal was still behaving as if the concrete were uncracked. Figure 4.15c shows the compressive surface strain of the concrete at knee J (see Plate 4.6) which proved to be considerably less than predicted and also indicated that joint J was behaving as if fully continuous.

Surface diagonal strains on the concrete were measured with gauges 1N to 3N and 1S to 3S, mounted on each face of the beam as shown in Figure 4.13. The structure was loaded in five approximately equal load increments, and the portal is orientated east-west. Since load increments 1 to 3 were placed on the northern deck span, the torsional strain would have reached a maximum with the application of the third load increment. It would be anticipated, therefore, that diagonal strains due to torsion would increase up to load increment 3 and that thereafter as loads were applied to the southern span, the effect of torsion would reduce to zero at full load and that strains due to pure shear would dominate. This pattern is evident in Figure 4.15d, where the measured diagonal strains at gauge 2N have been compared with the predicted behaviour. Considering the limited resolution of the Demec strain gauge (5×10^{-6}), the measured strains follow the calculated trend very well and were also less than predicted.

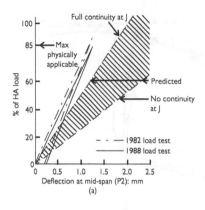

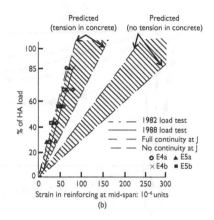

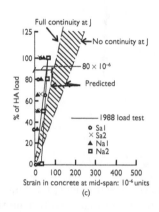

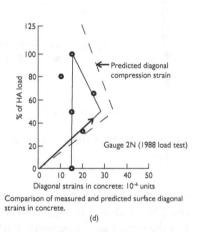

Figure 4.15 Comparison of predicted (calculated) strains in the portal frame with values measured in the two full-scale loading tests of 1982 and 1988, under 100% NA loading (=81% of HA loading).

The predicted movements and strains shown in Figure 4.15 were based on a value of the elastic modulus derived from measurements made on cores drilled from the structure. This value was 18 GPa. Without two exceptions, measured movements and strains, both in 1982 and 1988, were less than predicted by between 21 and 33% of the measured value. The two exceptions were the in-plane rotation of joint J, where the measured values averaged the predicted value for full continuity of the joint, and the strain in the reinforcing at midspan of the beam, where the predicted strain was close to the lower limit of the predicted strain, assuming that the concrete took full tension, but 6.3% less than the prediction of no tension in the concrete. This means that the actual elastic modulus of the concrete was not 18 GPa, but between 22 and 24 GPa. In the measurements on cores, for which 18 GPa had been the mean, individual measurements, all on concrete damaged by AAR, had varied from 16 to 22 GPa.

This series of load tests is particularly interesting as, when the portal frames were designed in 1963, it was decided to check the design by means of a test on a large scale (one fifth full size) reinforced concrete model. The main aim of the model was to test the complex monolithic intersection of the portal column and lower beam with the deck slabs. It is possible, therefore, to compare the behaviour of the model (Ockleston, 1963) with that of the AAR-deteriorated prototype 20 years later. Plate 4.13 (re-photographed from the original 1963 photo) shows the model set up in its test frame (on right) with the foot of the column in the foreground and the beam-column-deck slab intersection and the load activator in the background. (The senior author is second from the left in the photograph.) Some of the results obtained in the model test are shown in Figure 4.16. (In the figure, the zeros have been displaced by 0.5 mm

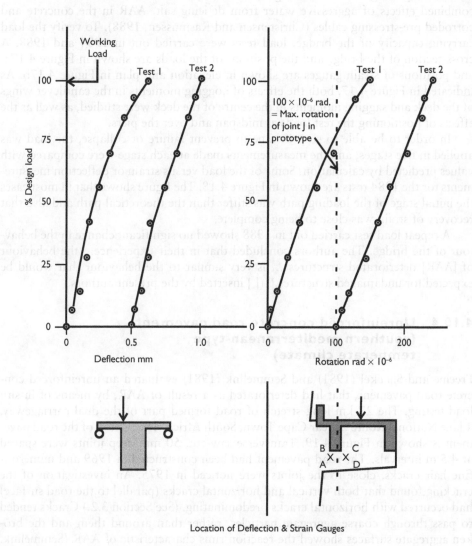

Location of Deflection & Strain Gauges

Figure 4.16 Measurements made on 1963 model of part of portal frame.

and 100 rad $\times$ 10^{-6} to the right, respectively, for test 2.) The model test results cannot be compared directly with those for the full-scale load test, but it is interesting to see that both the measured deflections and the rotations in the model were of the same order of magnitude as those measured on the AAR-deteriorated prototype frame. The measurements on the model also show that the degree of linearity of the prototype's deflection and strain reaction to load as well as their recoverability in 1982/88 were excellent, in fact, as good as those of the model in the 1963 test.

4.10.3   Motorway bridge
(northern cold-temperate climate)

A motorway bridge near Copenhagen, Denmark, had deteriorated as a result of the combined effects of aggressive water from de-icing salt, AAR in the concrete and corroded pre-stressing cables (Christensen and Rasmussen, 1988). To verify the load carrying capacity of the bridge, load tests were carried out in 1984 and 1988. A cross-section of the bridge and the positions of the loads are shown in Figure 4.17a, and positions of strain gauges are shown in elevation and plan in Figure 4.17b. As indicated in Figure 4.17, both the effects of hogging moments in the cantilever wings of the deck and sagging moments in the centre of the deck were studied, as well as the effects of positioning the test loads at midspan and over the piers.

In order to be able to stop in time to prevent failure or collapse, the load was applied in five stages, and the measurements made at each stage were compared with values predicted by calculation. Some of the load versus strain or deflection measurements for the 1984 tests are shown in Figure 4.18. The figure shows that in most cases the initial stage of the loading path was stiffer than the theoretical path and also that recovery of strain was close to being complete.

A repeat load test carried out in 1988 showed no significant change in the behaviour of the bridge. The authors concluded that in their experience: "the behaviour of [AAR] deteriorated structures ... is very similar to the behaviour that would be expected for undamaged structures." ([ ] inserted by the present authors.)

4.10.4   Unreinforced concrete road pavement
(southern mediterranean-type
temperate climate)

Freeme and Shackel (1981) and Semmelink (1981) evaluated an unreinforced concrete road pavement, that had deteriorated as a result of AAR, by means of in situ load testing. The 27 km long stretch of road formed part of the dual carriageway, 4 lane National Route 2, near Cape Town, South Africa. The section of the road pavement is shown in Figure 4.19. Transverse saw-cut, 50 mm deep joints were spaced at 4.5 m intervals. The road pavement had been constructed in 1969 and numerous fine hair cracks, close to the joints were noticed in 1975. An investigation of the cracking found that both vertical and horizontal cracks (parallel to the road surface) had occurred with horizontal cracks predominating. (See Section 3.2.) Cracks tended to pass through coarse aggregate particles, rather than around them and the broken aggregate surfaces showed the reaction rims characteristic of AAR (Semmelink, 1981). Plate 4.14, (photographed from the original Freeme and Shackel paper), shows

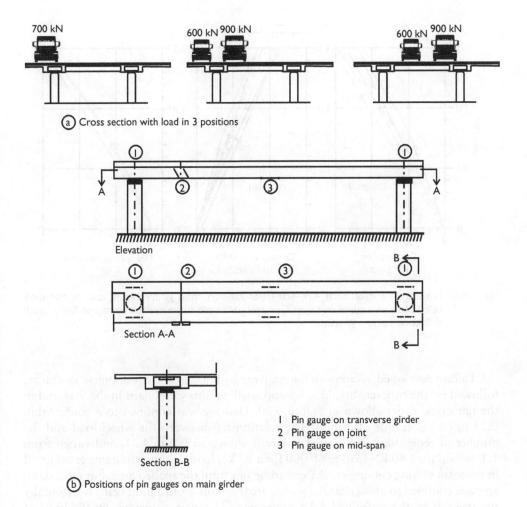

700 kN

600 kN 900 kN

600 kN 900 kN

(a) Cross section with load in 3 positions

Elevation

Section A-A

1 Pin gauge on transverse girder
2 Pin gauge on joint
3 Pin gauge on mid-span

Section B-B

(b) Positions of pin gauges on main girder

Figure 4.17 Test load positions and locations of strain and deflection measuring pins on structure.

cracking close to a transverse joint, revealed by a longitudinal saw-cut. The vertical crack close to the transverse joint (upper right corner) is clearly shown together with a system of three horizontal cracks (lower foreground). Plate 4.15 shows cracking that occurred close to the transverse joints, which is also shown diagrammatically by Figure 4.20.

In 1979 it was decided to investigate the effect of the AAR on remaining pavement life by loading selected areas of the pavement by means of a heavy vehicle simulator, or HVS. An HVS consists of a specially constructed chassis, mounting a wheel that can be loaded to various degrees by means of a hydraulic cylinder that reacts against the heavily ballasted chassis. The loaded wheel can be trafficked back and forth along the test strip of pavement. Tests were carried out in two areas, applying wheel loads of 40, 60 and 95 kN and continuing applications until the pavement failed. The maximum legal single wheel load was 40 kN at the time.

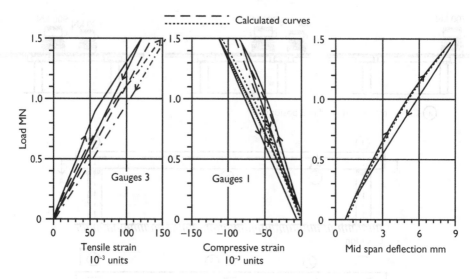

Figure 4.18 Load/strain curves: Span 409–410 (over railway), main girder No. 4, load at mid-span. Left: Load/strain curve mid-span, Centre: Load/strain curve over columns, and Right: Load/ deflection curve mid-span.

Failure developed adjacent to the transverse joints, by the development of cracks, followed by the concrete breaking up and spalling into small slabs in the way and in the numerical order shown in Figure 4.20. Damage was confined to a zone within 0.25 m on either side of a joint. The relationship between the wheel load and the number of repetitions that caused failure is shown in Figure 4.21, and varied from 1.1 million for a 40 kN load to 60 000 for a 95 kN load. Ultimately, damage occurred in the form of long curving cracks spreading out from the spalled area, but most damage was confined to areas near the joints, areas remote from a joint being horizontally pre-stressed by the restrained AAR expansion. This was estimated, on the basis of strain release tests on cores, to amount to as much as 4 MPa horizontal stress in concrete which had a mean core strength in 1979 of 33 MPa and an estimated flexural tensile strength of 4.5 MPa. (Also see Section 3.8.2.)

Hence, away from the discontinuities represented by the joints, the concrete continued to be strong and serviceable despite the occurrence of AAR. The estimated residual life of the pavement in 1981, under legal wheel loading, was 5 to 12 years over the most heavily trafficked sections, giving a total life, before rehabilitation was predicted to be required, of 17 to 24 years, as compared with the design life of 25 years.

4.10.5 Underground mass concrete plug

Pneumatic tools and machinery are extensively used in underground mining, usually in conjunction with a reservoir or receiver for the compressed air supply, that often takes the form of a chamber or blind drive excavated in rock and sealed by

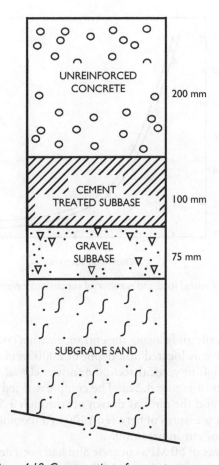

UNREINFORCED
CONCRETE   200 mm

CEMENT
TREATED SUBBASE   100 mm

GRAVEL
SUBBASE   75 mm

SUBGRADE SAND

Figure 4.19 Cross-section of concrete pavement.

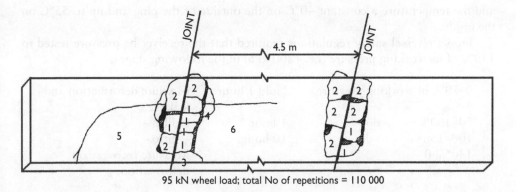

95 kN wheel load; total No of repetitions = 110 000

Figure 4.20 Progression of spalling and cracking in concrete pavement.

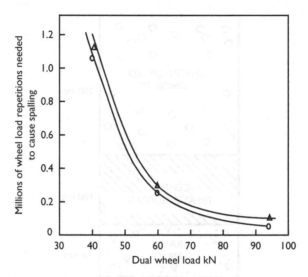

Figure 4.21 Effect of wheel load and number of repetitions on spalling of concrete.

gunite-lining the rock walls and closing the entrance with a concrete plug. In this case the reservoir or receiver was located at a depth of 2000 m in a gold mine, measured 3.5 m wide by 3.3 m high in vertical section and was closed by a 1.5 m thick plain concrete plug, as shown in Figure 4.22a. The compressor fed the receiver through a 250 mm diameter pipe and the air was drawn off through a 500 mm diameter pipe. The receiver operated at pressures of between 1400 kPa (compressor cut-out pressure) and 600 kPa (compressor cut-in pressure).

The concrete plug was of 50 MPa concrete and had been designed and constructed without any thought of the possibility of AAR, although the aggregate used consisted of Witwatersrand quartzite. However, in 1981 when the plug was designed, it was not generally known that concrete containing Witwatersrand quartzite could be subject to attack by AAR. The air on both sides of the plug was at 100% relative humidity and the temperature a constant 40°C on the outside of the plug and up to 55°C on the inside.

Pressure vessel safety regulations required that the receiver be pressure-tested to 130% of its working pressure (i.e. 1800 kPa) in the following stages:

0–70% of working pressure.		Hold 1 hour	Measure deformation and leakage.
70–100%	– do –	1 hour	– ditto –
100–130%	– do –	10 hours	– ditto –
130%–0			Measure recovery of deformation.

A request was received from the mine management to plan and carry out the proof test. A system of dial gauges and LVDTs was designed, supported on a rigid

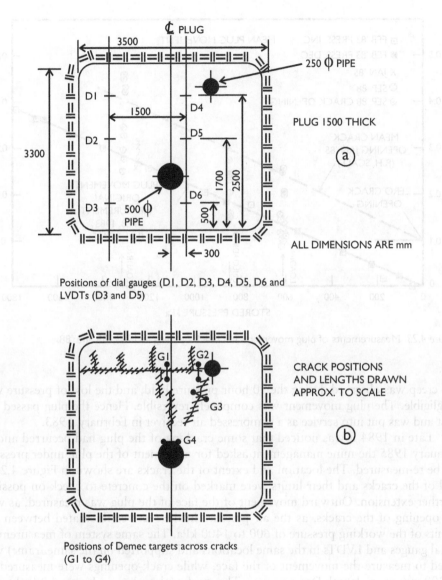

Positions of dial gauges (D1, D2, D3, D4, D5, D6 and
LVDTs (D3 and D5)

Positions of Demec targets across cracks
(G1 to G4)

Figure 4.22 Layout of measuring points on plug face.

framework constructed of welded steel scaffold poles that stood free of the plug and
were wedged against the rock walls, floor and roof. Both mechanical dial gauges and
LVDTs were used as it was not certain if electrics would work satisfactorily in the
100% humidity underground.

Because of the explosive nature of compressed air, the test pressurization was carried
out using water as the pressurized fluid. The results of the test are shown in Figure 4.23.
Average outward movement of the plug face, measured by 0.01 mm dial gauges and
LVDTs calibrated to 0.01 mm (see Figure 4.22a for location), was a very small 0.4 mm,

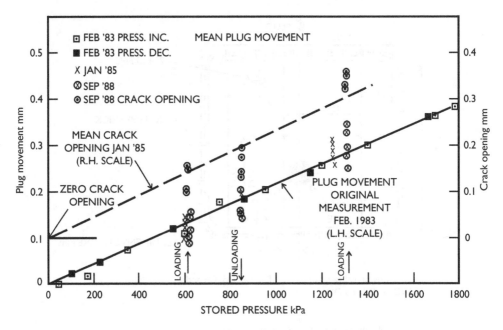

Figure 4.23 Measurements of plug movement and crack opening in 1983, '85 and '88.

no creep was observed during the 10 hour pressure hold, and the loss of pressure was negligible. The plug movement was completely reversible. Hence the plug passed the test and was put into service as a compressed air receiver in February, 1983.

Late in 1984 it was noticed that some cracking of the plug had occurred and in January 1985 the mine management asked for movement of the plug under pressure to be re-measured. The location and extent of the cracks are shown in Figure 4.22b. All of the cracks and their limits were marked on the concrete to check on possible further extension. Outward movement of the face of the plug was measured, as well as opening of the cracks, as the air pressure in the receiver fluctuated between the limits of the working pressure of 600 to 1400 kPa. The same system of measurement (dial gauges and LVDTs in the same locations and supported on the same frame) was used to measure the movement of the face, while crack-openings were measured by a 200 mm gauge length Demec gauge. The results (also given in Figure 4.23) showed close to the same relationship between air pressure and plug movement found in the original proof-test, although the measurements were somewhat more randomly variable. There was no indication that the plug was moving progressively. The cracks opened and closed in sympathy with the face movement, following a line parallel to that for the face movement (see Figure 4.23) and pressure losses were negligible. The air receiver was allowed to continue in service without any attempt to repair the cracks, but it was decided to repeat the measurements in 1986.

As it happened, these measurements were not made until September 1988 when, fortunately, they showed no deterioration of the condition of the plug. As shown in Figure 4.23, the September 1988 measurements were almost identical with those made in January 1985, 3.7 years earlier and the cracks had not extended. (The experimental points for crack-opening in 1985 have been omitted from Figure 4.23 to avoid graphical congestion.)

In 1990, the air receiver was decommissioned because the sector of the mine it served had been worked out. The mine was asked to provide a sample of the concrete for examination and, eventually, a piece with a mass of about 5 kg was provided that had been broken out of the plug using a jack-hammer. This showed all the visual signs of AAR attack, including reaction halos around the large particles of aggregate, cracks through the aggregate particles, etc. This was the first time that it was confirmed that the plug had been subject to attack by AAR, a suspicion that had been growing since the cracking was noticed in 1984. In retrospect, the plug acted as a giant accelerated AAR specimen (see Section 2.3.2).

A number of points are illustrated by this case history:

- All of the necessary conditions for AAR to develop were present, i.e. an AAR-susceptible aggregate, a constant supply of free moisture, a high and constant temperature and, with a required strength of 50 MPa in difficult mixing and placing conditions, (probably) a cement-rich concrete mix.
- The initial cracking may have resulted from shrinkage, but any shrinkage must rapidly have been reversed by expansion caused by AAR as well as compressive creep in the highly stressed rock walls (overburden stress of 56 MPa) confining the sides of the plug. The cracks obviously did not open into the void of the receiver because leakage of air remained negligible. The directions of the major principal stresses on the plug were probably vertical and horizontal, resulting from closure of the walls of the rock excavation as well as restrained swelling of the concrete caused by AAR. If tension cracks parallel to the free faces of the plug had occurred, they did not lead to any significant pressure losses.
- The measurements show that the strength and elastic modulus of the concrete did not deteriorate to any significant degree. Although it had been attacked by AAR, the expansion caused by AAR had been restrained by the confining rock walls.

A recent survey was made of the condition of concrete linings of tunnels in Switzerland, aged between 19 and 82 years and serving as railway, road or maintenance access in hydroelectric plants (Leemann, et al., 2004). About 60% of the concrete aggregates used are known to have a potential for AAR, the relative humidity in the tunnels can vary between 40 and 80% and the temperature may be as high as 40°C. Some of the newer (19 to 44 year-old) concretes were found to exhibit signs of AAR cracking, but no substantial damage had occurred.

In 2008 an inspection was made of a 1.25 m diameter by 1500 m long penstock decant outfall pipe serving a large tailings dam in South Africa. The pipeline was 30 years old at the time and was constructed of spun concrete pipes with Witwatersrand quartzite aggregate. In 1978 when the pipes were made, it had not yet been recognized that a combination of high cement content and Witwatersrand quartzite aggregate could (possibly) result in AAR. The pipe showed no signs of AAR damage, although some cracking associated with the superimposed load of 40 m of tailings was observed. On this limited evidence, AAR may not be a problem in concrete used underground and which is confined by compressive loading.

4.10.6 Industrial structural pavement

A factory manufacturing diesel engines and gearboxes had an outdoor unloading area where palletized components were loaded and unloaded from trucks by means of fork-lifts. The area was paved with a concrete slab that had been cast on a compacted soil base. The slab was a nominal 150 mm thick and was lightly reinforced with a layer of 150 mm square weld-mesh of 13 mm (0.5 inch) diameter bars. The mesh should have been located at mid-depth of the slab. The slab was divided into 5 m square panels by 15 mm deep saw-cut joints.

In 1981, ten years after construction, the surface of the slab had developed a network of randomly distributed fine hairline cracks. An investigation of the cracking was undertaken that involved the examination of cores drilled from points selected at random over two areas of the slab. During the coring, it was discovered that the concrete varied in thickness from 112 to 159 mm with a mean thickness, based on 13 cores, of 136 mm. During placing, the mesh had been displaced downwards in the concrete and the bottom cover varied from 5 to 45 mm. Examination of the cores showed that the concrete, made with AAR-susceptible Witwatersrand quartzite aggregate, was in the early stages of developing AAR.

It was decided to carry out a test loading to assess the structural behaviour of the slab.

The most severe working loads would arise from the passage of a fully loaded fork-lift carrying a maximum load of 20 kN, rolling diagonally over the junction of four slab panels at 45° to the joints. Tensile strains in the underside of the slab would be limited in their crack-forming effects by the mesh reinforcing which should prevent tension cracks originating at the underside from penetrating to the surface. The surface, however, was already cracked, although when examined, the cracks were all closed.

Electric resistance strain gauges (compensated for temperature changes), with a gauge length of 50 mm, were glued to the concrete surface in the positions shown in Figure 4.24. The loading vehicle was a standard 2 tonne fork-lift carrying its maximum load of two 1000 kg cases of gearbox castings. Including the tare load, this gave a wheel load of 20 kN. The loaded wheel was brought onto the slab in a diagonal direction, rolling from A to F (in Figure 4.24) and then back to A. The distance of 200 mm between the rows of gauges was wide enough to allow the tyre of the loading wheel to pass without touching any of the gauges. (The measuring gauges were orientated parallel to line AF.)

Static strain measurements were made with the wheel over points A to F, and an ultra violet light (UV) recorder was then used to take continuous recordings of strain at points 1, 3, 4 and 6 as the wheel rolled from A to F and back to A. Sets of measurements were made in three areas where the slab thickness at the junction of the four panels had been measured as:

area 1–112 mm
area 11–130 mm
area 111–150 mm

Figure 4.25 shows a selection of the static strain measurements. At point I, with the load at A, tensile strains were negligible for areas I and II, and perhaps surprisingly, were

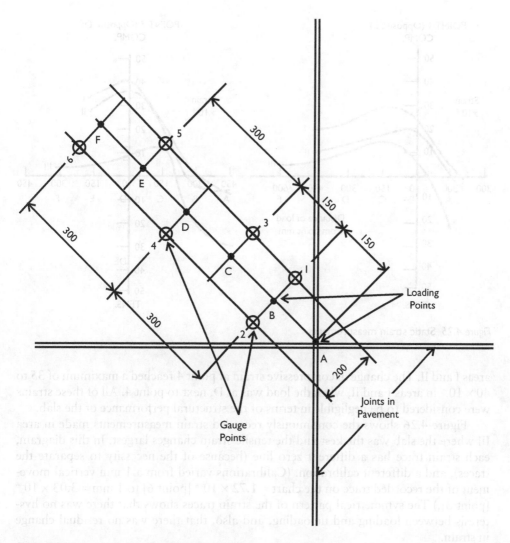

Figure 4.24 Layout of strain gauges for load test.

greatest where the slab was thickest, in area III. However, it must be remembered that the slab was probably in horizontal compression as a result of the AAR, a likelihood that was not appreciated in 1981 when the test loading was done. The strains that were being measured were very likely to have been relatively small changes to a larger bi-axial horizontal compressive strain field, caused by the AAR. As the wheel rolled towards point B from A, the strain at point I changed to an additional compressive strain that reached a maximum when the wheel was over point C. The largest measured tensile strain of 20×10^{-6} occurred in area III and the largest compressive strain occurred in area I.

At point 4, the changes in strain were fairly similar. With the wheel at A, the change in strain at point 4 was a maximum of 30×10^{-6} tension in area III, and almost zero in

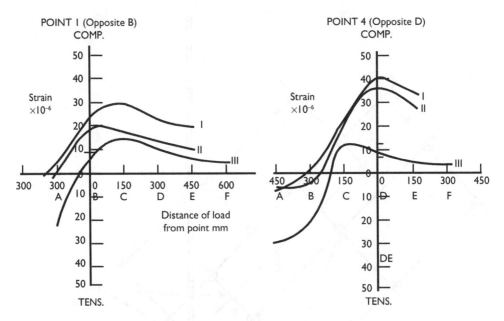

Figure 4.25 Static strain measurements.

areas I and II. The change in compressive strain at point 4 reached a maximum of 35 to 40×10^{-6} in areas I and II, when the load was at D, next to point 4. All of these strains were considered to be negligible in terms of the structural performance of the slab.

Figure 4.26 shows the continuously recorded strain measurements made in area III where the slab was thickest and the tensile strain changes largest. In this diagram, each strain trace has a different zero line (because of the necessity to separate the traces), and a different calibration. (Calibrations varied from a 1 mm vertical move-ment of the recorded trace on the chart = 1.72×10^{-6} [point 6] to 1 mm = 3.03×10^{-6} [point 3].) The symmetrical pattern of the strain traces shows that there was no hys-teresis between loading and unloading, and also, that there was no residual change in strain.

It was concluded that the slab was behaving perfectly elastically under the pas-sage of the maximum working load and that the structural capability of the slab had not been compromised either by the under-thickness of the slab, the misplacement of the steel mesh, or by the effects of the AAR. In retrospect, any biaxial compression developed by the AAR must have improved the performance of the slab under load.

Notwithstanding the results of the test, the owners decided to take almost imme-diate measures to ensure against deterioration of the slab. Three months later, during the industrial shutdown over Christmas, the slab was overlaid with a 50 mm thick layer of a patented material known as "Salviacim". A bitumen prime was applied to the surface of the concrete, followed by a high void content 50 mm thick layer of fine, single sized gravel, hot premixed with a bitumen binder. The voids in the bitumen-bound layer were then filled by vibrating a plasticized cement grout into the voids, to give a dense, impervious, though flexible surface layer.

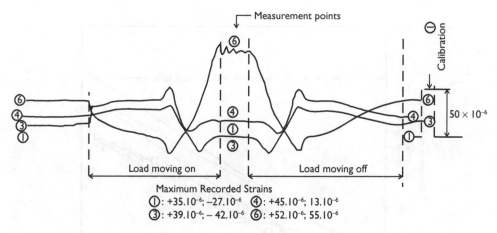

Maximum Recorded Strains
①: +35.10⁻⁶; –27.10⁻⁶ ④: +45.10⁻⁶; 13.10⁻⁶
③: +39.10⁻⁶; – 42.10⁻⁶ ⑥: +52.10⁻⁶; 55.10⁻⁶

Figure 4.26 Strain measurements under rolling wheel.

Unfortunately, because the factory was manufacturing military equipment, the usual mindless military secrecy prevented any follow-up on the post-1981 performance of the slab.

4.11   ROUTINE PERIODIC TEST LOADING OF COMPLETE STRUCTURES

4.11.1   Loading jetty over sea (southern moist tropical coastal climate)

Construction of a 5.76 km long jetty, serving as an outloading port facility for raw sugar at Lucinda, North Queensland, Australia, was started in 1977 and completed in 1979. The jetty supports a conveyor belt and its weather-proof housing, as well as a roadway providing access for maintenance of the conveyor and to the off-shore loading wharf. The 3.6 m wide roadway spans are 20 m long with 8.2 m wide passing locations at 6 points along the jetty. The roadway consists of 6 parallel pre-stressed concrete box girders (14 at passing points) post tensioned transversely together, each girder measuring 660 mm deep by 596 mm wide. Cracks appeared in the girders a few years after completion, the cause being diagnosed as AAR. A full investigation into the deterioration of the roadway was undertaken in 1986 (Holm, Idorn and Braestrup, 1988). Plate 4.16 shows the cracked undersides of the beam (re-photographed from Holm, et al.). The main crack direction is parallel to the length of the beam (and the photograph).

Prior to 1986, in 1985, a program for test loading selected spans at regular intervals had been introduced. Loads are applied by means of a specially designed trailer, and deflections are measured relative to the independently supported conveyor gallery. The spans are loaded incrementally up to the full design load. Figure 4.27 shows some typical plots of load versus deflection for three of the spans. There is some

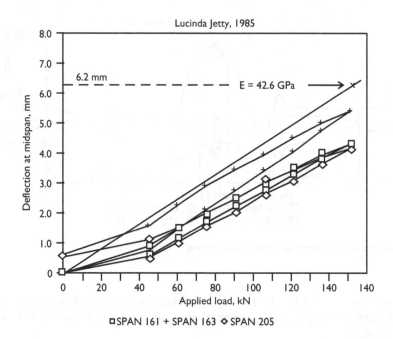

Figure 4.27 Mid-span load-deflection curves from 1985 load-tests on Lucinda jetty. The deflection calculated by elastic analysis is indicated by an x.

hysteresis between loading and unloading, but very little residual deflection. In Figure 4.27, span 163 shows the largest deflection of 5.2 mm. A core taken from this span gave a compressive strength of 62 MPa, an elastic modulus of 42.6 GPa and a calculated deflection of 6.2 mm.

Since 1993, load tests have been carried out yearly (Carse, 2004) and Figure 4.28 shows the variation in elastic moduli, calculated from the measured midspan deflections up to 2002, for the same spans that feature in Figure 4.27. In the 17 years after 1985, the calculated modulus reduced by about 10% from 55 GPa to 50 GPa. The conclusions in 2004 (Carse) were:

> "The structure is stable in the longitudinal direction, with no measurable expansion due to A[A]R occurring...." "All tested spans indicated a normal response [to loading] was obtained with no significant change in long term stiffness."

4.11.2   Bridges on highway (north temperate climate)

All of the 224 bridges on the A26 highway in France, built in the 1970s, are affected to varying degrees by AAR and/or sulfate attack on the concrete. As part of a detailed surveillance system for these bridges, routine loading tests are carried out at regular intervals, the first loading test having been performed in 1980. Baillemont et al. (2000) report on the results of a typical loading test. The paper is rather difficult to follow, but our understanding is that the load is applied by means of a 48 T truck

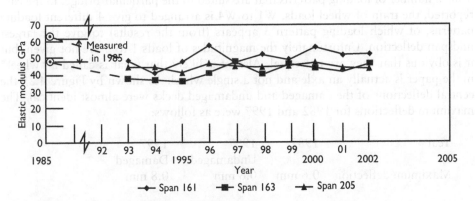

Figure 4.28 Variation in elastic moduli from 1985 to 2002.

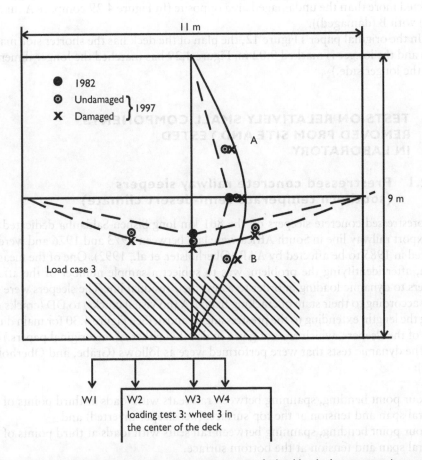

Figure 4.29 Plan of test bridge decks showing arrangement of wheel loads that gave maximum deflection. The area of deck damaged by AAR is cross-hatched and deflection profiles for the first load test in 1982 and similar results for what appear to be the 1997 loading of identical undamaged and damaged decks are superimposed on the plan.

using a number of loading patterns that are suited to the particular bridge. In the case reported, the train of wheel loads, W1 to W4 is arranged to give 4 different loading patterns, of which loading pattern 3 appears (from the results) to give the largest midspan deflection. Unfortunately the magnitudes of loads 1 to 4 are not given, but it is obvious that they are not equal. Also, it is likely that what is called a "wheel" in the paper is actually an axle and not a single wheel. As shown by Figure 4.29 the central deflections of the damaged and undamaged decks were almost identical. The maximum deflections for 1982 and 1997 were as follows:

Year	1982	1997 Undamaged	1997 Damaged
Maximum deflection	0.6 mm	0.6 mm	0.8 mm

Hence in 15 years, the maximum deflection of the decks under the test loading has hardly changed, although Baillemont et al. note that the area of damaged deck deflected more than the undamaged area opposite (In Figure 4.29 compare A (undamaged) with B (damaged)).

(In the original paper's Figure 12, the plan of the deck has the shorter side marked 11 m and the longer is marked 9.02 m. Figure 4.29 has matched the longer dimension with the longer side.)

4.12　TESTS ON RELATIVELY SMALL COMPONENTS REMOVED FROM SITE AND TESTED IN LABORATORY

4.12.1　Prestressed concrete railway sleepers (southern temperate semi-desert climate)

The prestressed concrete sleepers on the 861 km long Sishen-Saldanha dedicated iron ore export railway line in South Africa were laid between 1973 and 1976 and were discovered in 1985 to be affected by AAR (Oberholster, et al., 1992). One of the measures taken, after identifying the problem, was to subject a sample number of the affected sleepers to dynamic loading in order to assess their residual life. The sleepers were classified according to their state of cracking from AA (no visual cracks) to DD (cracks right along the length, extending to the prestressing strands). (See Figure 4.30 for main dimensions of the sleepers, which are not given in detail in either of the original papers.)

The dynamic tests that were performed were as follows (Gräbe, and Oberholster, 2000):

- four point bending, spanning between rail seats with loads at third points of central span and tension at the top surface, (i.e. sleeper inverted) and
- four point bending, spanning between rail seats with loads at third points of central span and tension at the bottom surface.

In all tests, the sleepers were subjected to dynamic loading equivalent to 1.15 times the theoretical tensile cracking load of the sleeper for at least 5×10^6 cycles. This is equivalent to 5 years of service in the line with 260 kN axle loads. Of 12 tests

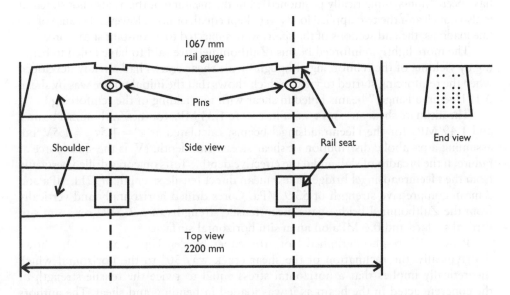

Figure 4.30 Views of prestressed concrete sleepers used on Sishen-Saldanha iron ore export rail line.

on sleepers in the worst cracked (DD) category, 10 sleepers endured 10×10^6 cycles of loading, and at completion, none of the cracks exceeded the specified maximum width of 0.1 mm. The testing program was completed in 1995.

The residual life of the sleepers was assessed as 0 to 2 years for class DD, 2 to 5 years for class C and 30 years for class AA. A program of in situ assessment and replacement of sleepers classed C and DD was instituted in 1995. Between 1995 and 1999, 37500 sleepers were replaced (approximately 4% of the total number) and this was expected to increase to 31000 (3%) per year in 2005. Simultaneously, a program of treating sleepers with silane to slow the AAR process was started. A 100% iso-butyltrimethoxysilane was specified for use. After clearing the ballast from the sides of each sleeper, the silane was sprayed on the top and side surfaces with a control-led rate of application. Between 1995 and 1999, 440 000 sleepers were treated with silane (of the approximately 1 million in the entire line, excluding passing points).

4.12.2   Beams sawn from flat slab bridges (northern cold-temperate climate)

Six beams were sawn from the decks of two 35 year-old viaducts in the Netherlands, located at and referred to as "Heemraadsingel" and "Zaltbommel" (den Uijl, et al., 2000). The beams were transported to Delft University where they were tested in 4-point bending. The cross-sections of the beams varied from 620 to 740 mm deep and the lengths were either 7.5 or 8.5 m. The sections were reinforced with longitudinal and transverse bars at the top and bottom, but there were no vertical links between the upper and lower reinforcing mats and hence the shearing strength had to be provided by the concrete. The spans over which they were tested are not stated. The loads are said to have been "non-symmetrically positioned", but the meaning of this is also not stated. It is also not clear if the two applied loads were kept equal, or not. However, because of the line loadings, the end sections of the spans were subjected to a constant shear force.

The more lightly reinforced beams (Zaltbommel) are said to have failed in bending by yielding of the reinforcing although "wide shear cracks had already developed when the reinforcing started to yield" which shows that the initial failure was by shear. All "Heemraadsingel" beams failed in shear with no yielding of the reinforcing.

The ultimate shear stresses τ_u at failure were 1.34 MPa for the Zaltbommel beams, and 1.49 MPa for the Heemraadsingel beams, calculated as $\tau_u = 1.5\tau_{av} = 1.5V_u/bd$, assuming a parabolic distribution of shear stress with depth. (V_u is the shear force at failure, b the breadth of slab and d the effective depth.) Tests on cores drilled vertically from the Heemraadsingel bridge gave a mean direct tensile strength of 1.11 MPa and a mean compressive strength of 53.9 MPa. Cores drilled horizontally and vertically from the Zaltbommel bridge gave mean tensile strengths of 1.2 MPa on an in situ vertical surface and 0.6 MPa on an in situ horizontal surface.

Plate 4.17, re-photographed from the paper by den Uijl et al. (2000), shows that typically, the inclination of the shear crack was 30° to the horizontal which theoretically implies that a horizontal stress equal to twice the tensile strength of the concrete acted in the beam as it was loaded in bending and shear. The authors conclude that:

• "The shear strength of a member without shear reinforcement that suffered from [AAR] can be estimated on the basis of the tensile strength in a vertical direction [i.e. on a horizontal plane], the tensile strength ratio and the prestress caused by [partially restrained AAR] expansion." (The tensile strength ratio is defined as the ratio of the tensile strength of the concrete normal to the shear crack to the tensile strength across a horizontal surface.)

Very importantly:

• "The average capacity in case of shear failure was 75% of the value that would have been expected when no [AAR] damage had occurred."

It follows that if the shear capacity of the AAR-damaged sawn-out beams had been 1.5 of the design loading, the factor of safety against failure under the design loading would have been 1.13. Knowing that the design loading on road bridges is physically very difficult to achieve (see Section 4.4.1), it is likely that in situ, with the horizontal stresses caused by restrained AAR expansion still acting, the shear capacity would have been adequate for safety.

4.12.3   Prestressed planks taken from road bridge (southern warm-temperate climate)

The six span road bridge in New South Wales, Australia, was built in 1989 with a deck consisting of precast prestressed steam-cured planks laid side by side and spanning between the piers (Shayan, et al., 2008). An investigation of cracking of the deck planks in 2001 resulted in the conclusion that this, and several similar bridges were deteriorating as a result of AAR or combined AAR and DEF (Delayed Ettringite Formation). In 2003, the deteriorated deck planks were removed when this bridge was repaired. To assess the remaining strength of the planks, four were selected, representing the range of deterioration visible on the bridge, and were taken to a laboratory for testing. As the planks were 11.8 m long and weighed 65 kN, each was cut in half to reduce the difficulty of handling them. Apart from tests on cores taken from the planks, four half-lengths (beams B1 to B4) were tested to failure in 4-point bending.

Figure 4.31 shows a cross-section of the planks and the layout and dimensions of the loading arrangement (not to scale). It also shows the moment versus midspan deflection curves for the four planks. Note that planks B1, B2 and B3 were not taken to their maximum, i.e. failure load. The moment on plank B4 stepped down at a deflection of 50 mm, but then recovered and at a deflection of 67 mm had surpassed the value at which it had faltered.

The design strength of the concrete was 40 MPa, and strengths and elastic moduli measured on cores taken from the planks were as set out below, together with maximum values, in the tests, of midspan bending moment (BM) and load (W_f):

		B1	B2	B3	B4
Compressive strength	MPa	70	38	67	69
Elastic modulus	GPa	28	15	26	27
Max. midspan BM	kNm	691	665	658	627
Maximum load	W_f kN	307	295	292	278
Load factor	LF = W_f/120kN	2.55	2.46	2.43	2.31

The loads were arranged (see Figure 4.31) to simulate a dual axle with wheel loads W. The design load is not given in the paper, but when one considers that maximum legal axle loads are usually in the range of 80 to 120 kN it will be seen that if the minimum load factor had been $(W_f/120$ kN), it would have been a very adequate 2.31 for beam B4.

Plate 4.18, re-photographed from a small print in the original paper, has been included to show that at the end of the bending tests, once the loads had been removed, the cracks closed completely and there is little evidence of concrete crushing. This indicates that the prestressing steel was still perfectly elastic under the maximum load applied and that there was a reserve of strength in the concrete. The complete recovery means that strictly, the planks did not fail and would probably have reached the same load if reloaded. This is not altogether surprising when one considers that the maximum deflection was only 70 mm in a span of 5350 mm (1 in 76, or 1.3%. 1 in 100 is usually regarded as acceptable). Thus the strength of the planks proved to be completely satisfactory. The authors still questioned the planks' long-term durability in service, due to potential residual lateral expansion of the concrete. However, because the planks were laid side by side and probably with the gap between planks solidly caulked, any potential lateral expansion would have been well restrained and unlikely to affect the strength of the planks to a significant degree.

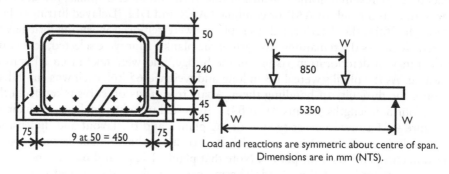

Load and reactions are symmetric about centre of span.
Dimensions are in mm (NTS).

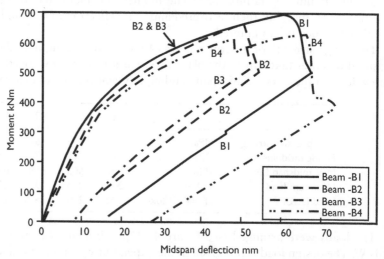

Figure 4.31 Moment-deflection curves for tests on AAR-deteriorated pre-stressed concrete planks recovered from a road bridge (Plank cross section and loading arrangement – not to scale-above.).

4.13   REVIEW AND CONCLUSIONS

It is interesting to see that most of the published results on full-scale loading tests on structures date from the early to mid-1980's and the late 1990's and early 2000's. In most cases these have been one-off tests, or tests that were repeated after an interval of a few years to check on the possibility that structural deterioration was progressing. However, continuing series of periodic full-scale loading tests have been made on some structures, notably the Lucinda jetty in Australia (Holm, et al., 1988, Carse, 2004 and the A26 bridges in France (Baillemont et al., 2000). The Lucinda tests have been carried out over a period of at least 15 to 16 years in a tropical marine environment with warm, wet, salt-laden conditions that are likely to have been the worst possible for the progression of AAR. The Lucinda jetty roadway, however, shows no continuing deterioration of structural capability. In fact, apart from its external appearance, it has not deteriorated significantly at all. Similar comments seem to apply to the A 26 highway bridges in France despite exposure to rigorous climatic conditions. It would be nice to think that engineers have noted the conclusions drawn from these early and continuing experiences, including all eleven cases discussed in Chapter 4 and have learned from them that AAR (the so-called "concrete cancer") need not be a terminal ailment that necessarily ends in early demolition and rebuilding, but simply an additional repair and maintenance problem to add to the list of repair and maintenance problems that beset all exposed concrete structures.

In 1988, Braestrup and Holm, after a survey of their extensive experience with AAR-affected structures, concluded as follows:

- "The deflections of ASR-affected structures under 80–85% of service load may be predicted by standard elastic analysis, but the use of material properties obtained by testing of extracted cores will give conservative values. The response may not be much different from that of sound concrete. No significant creep deformations are observed, and the short term residual deflection is 15–20%", and
- "The load-carrying behaviour of structures suffering from [AAR] is generally much better than would be expected from the material properties measured on test specimens or cores. The unrealistic stiffness data from test samples are due to the fact that the specimens will have developed cracks perpendicular to the applied load, which is not [necessarily] the case in the structure. Indications are that material degradation due to [AAR] only affects [unconfined] concrete which has cracked and expanded, whereas the effect of [AAR] on concrete which is confined by reinforcement or compressive stresses is negligible or even beneficial".

Two of the present authors (Blight, Alexander, et al., 1989) concluded at follows:

- "... deterioration of concrete as a result of alkali-aggregate reaction may be alarming in appearance but is not necessarily structurally dangerous. Moreover, structures that have deteriorated behave predictably and their elastic behaviour can be predicted on the basis of laboratory tests on cores taken from the structure.

   The major effect of alkali-aggregate deterioration appears to be to [slightly] increase the deformation of a structure by reducing the elastic modulus of the concrete. With normal design practice where the design load usually far exceeds loads actually applied to a structure, safety appears not to be a serious problem. (With certain other classes of structure, where design loads can be determined with greater certainty and where the structures carry more of their design load, [e.g. liquid containing structures] this conclusion may not be valid.)"

It is believed that the above conclusions are as valid in 2010 as they were when they were first drawn.

It is also important to note the more specific and detailed conclusions of the British Institution of Structural Engineers (1992), extracts of which are quoted below. In these quotations, the more specific "ASR" has been changed to the more general AAR:

BEAMS

"AAR does not have a significant effect on the flexural strength of reinforced concrete beams provided that the free expansion does not exceed about 6 mm/m [6×10^{-3}]. At higher expansions strength reductions of up to 25% have been observed. Flexural strength can be assessed by using the uniaxial compressive strength of the AAR affected concrete in the conventional models of behaviour with due consideration of delaminations in the structure". [Delaminations have been observed to develop between the surface layer of steel and the concrete cover.]

"Tests have shown no significant decrease in shear capacity as a result of AAR if at least 0.2% links are present – indeed, an enhancement of strength of up to 47% has been observed in some tests. Included in the tests were beams with anchorages to main reinforcement as small as 3.4 times the bar diameter. This observed good behaviour in shear may arise from the self-prestressing effects of the restrained AAR expansion. It is advised that a lower bound value of the prestressing effect be used at the ultimate limit state. Doubts have been raised as to whether this prestress is maintained. However, data are now available which show that at least 40% of the AAR induced prestress is present after two years in drying conditions. The test evidence from beams without links is conflicting with some showing an increase in shear capacity and some showing a decrease of up to 20%.

It is suggested that the shear capacity of an AAR-affected beam can be estimated by considering the influence of the compressive stress resulting from the prestressing effect of the restraint. The mechanical properties of the AAR-affected concrete should be used in making this estimate." [Information on the prestressing effect of restrained AAR expansion is given in Section 3.8.]

"Repeated loading tests on reinforced concrete beams have demonstrated that AAR does not reduce the fatigue life, probably because the reinforcement stress range is reduced by the prestressing effect of the restrained AAR expansion.

Tests on post-tensioned beams and pre-tensioned prestressed concrete beams have led to similar conclusions to those for reinforced concrete beams."

COLUMNS

"AAR can have three effects on column behaviour:

- The concrete compressive strength can be reduced. The effects of the concrete compressive strength reduction can be allowed for by using the appropriate strength in the conventional models of behaviour..."
- "A second effect of AAR on the behaviour of columns is that delamination can occur in the plane of the reinforcement so that the concrete cover may not be effective in resisting compression..."
- "The compressive strain induced in the concrete as a result of the restraint to the AAR compression by the reinforcement could mean that, under load, the concrete could crush prior to the main steel yielding in compression. However, such crushing is extremely unlikely to occur in practice..."

SLABS

"In addition to the flexural and flexural shear failure modes, a slab can fail in punching shear. Tests have shown no significant reduction in punching shear strength for free expansions of up to 6 mm/m [6×10^{-3}]. Furthermore, AAR increases the ductility of a punching shear failure. For free expansions in excess of 6 mm/m, delamination in the plane of the reinforcement effectively divides a slab into three layers if there are no links tying together the top and bottom reinforcement. In such cases strength reductions of up to 30% can result. If there is significant surface cracking of a slab due to AAR, it is recommended that possible delamination should be checked by coring and petrographic inspection of the cores for sub-parallel cracking."

"The effects of AAR on punching shear capacity can be evaluated using conventional models of behaviour for prestressed concrete slabs. The appropriate compressive and tensile strengths of the AAR-affected concrete are used, and the precompression, due to the restraint to AAR expansion provided by the reinforcement, is allowed for by calculating the decompression load in the same way as for a conventional prestressed concrete slab."

BOND

"Anchorage bond and lapped bar tests... with both plain and ribbed bars have shown that AAR up to a free expansion of 4 mm/m [4×10^{-3}] does not have a significant effect on either type of bar when restrained by links and/or a substantial thickness of cover concrete of the order of four times the bar diameter. However, the bond strength of bars not restrained by links and with a cover of the order of 1.5 times the bar diameter is reduced by up to 50%. The reduction is proportional to the reduction in splitting tensile strength of the AAR-affected concrete."

TORSION AND BEARING

... "torsional shear capacity is more dependent that flexural shear capacity on the anchorage of links. This is because of the greater tendency for the cover concrete to spall under torsional shear.

No testing has been performed on the effects of AAR on bearing strength. At the edges of corbels and local bearing details the failure mode is by splitting, so that the concrete tensile strength reduction is relevant. In respect of bearing under bends in reinforcement, the tensile strength reduction is also relevant when the cover is small, say less than four times the bar diameter. In such situations, if there are no links or secondary reinforcement to restrain a splitting failure mode, allowable bearing stresses should be reduced in proportion to the reduced concrete splitting tensile strength."

It is obvious that once AAR damage has been diagnosed, it should be repaired as soon as possible. It must be recognized from the outset that repair is unlikely to be a once-off process. However, as will be seen from Chapter 5, many tried and tested repair methods are available that can extend the life of an AAR-damaged structure to the extent of its original design life and beyond. Even a completely undamaged structure requires continuing preventive maintenance, which involves periodic repair procedures. Necessary periodic repair of AAR-damaged structures may prove to be somewhat more onerous, but is simply the same process. Chapter 5 will describe a variety of tried, tested and successful repair strategies and techniques that can extend the working lives of AAR-damaged structures.

REFERENCES

Akashi, T, Amasaki, S, Takagi, N & Tomita, M 1986, 'The estimate for deterioration due to alkali-aggregate reaction by ultrasonic methods', *7th Int. Conf. on AAR in Concrete*, Ottawa, Canada, pp. 183–187.

Amasaki, S & Takagi, N 1989, 'The estimate for deterioration due to alkali-silica reaction by ultrasonic spectroscopy', *8th Int. Conf. on AAR in Concrete*, Kyoto, Japan, pp. 839–844.

Baillemont, G, Delaby, JB, Brouxel, M & Rémy, P 2000, 'Diagnosis, treatment and monitoring of a bridge damaged by AAR', *11th Int. Conf. on AAR.*, Quebec City, Canada, pp. 1099–1108.

Blight, GE 2006, 'Assessing loads on silos and other bulk storage structures', *Taylor and Francis/Balkema*, Leiden, The Netherlands.

Blight, GE & Alexander, MG 1982, 'Evaluating reinforced concrete structures affected by alkali aggregate reaction', *Int. Symp. on Re-evaluation of Concrete Structures*, Danish Concrete Institute, Copenhagen, Denmark, pp. 309–317.

Blight, GE, Alexander, MG, Ralph, TK & Lewis, BA 1989, 'Effect of alkali-aggregate reaction on the performance of a reinforced concrete structure over a six-year period', *Magazine of Concrete Research*, vol. 41, no. 147, pp. 69–77.

Blight, GE & Garstang, A 1987, 'Strains measured in a 7500 T sugar silo', *Bulk Solids Handling*, vol. 7, no. 4, pp. 573–582.

Blight, GE, Stewart, JA & Papendorf, GHW 1976, 'Deflection characteristics of an asphalt-paved steel bridge deck under vehicular loading', *Proc. 1976 Annual Meeting, Association of Asphalt Paving Technologists*, New Orleans, U.S.A., pp. 199–225.

Blockley, D, (Ed.) 1995. *Engineering Safety*, McGraw-Hill, Maidenhead, U.K.

Braestrup, MW & Holm, J 1988, 'Structural effects of alkali-silica reactions in concrete', *Int. Symp. on Re-evaluation of Concrete Structures, Danish Concrete Institute*, Copenhagen, Denmark, pp. 79–89.

British Institution of Structural Engineers, 1992, 'Structural effects of alkali-silica reaction', *The Institution*, London, U.K.

British Standards Institution, 1954, *Girder bridges*, BS 153, Part 3 A: Loads and Stresses, The Institution, London, U.K.

British Standards Institution, 1995, *Structural use of concrete*, BS 8110, Part 2, Code of Practice for Special Circumstances, The Institution, London, U.K.

Canadian Dam Safety Association, 1995, *Dam safety guidelines*, The Association, Edmonton, Canada.

Carse, A 2004, 'Review of the present condition of the Lucinda bulk sugar terminal at Lucinda in North Queensland, Australia', *12th Int. Conf. on AAR in Concrete*, Beijing, China, pp. 1025–1034.

Christensen, HH & Rasmussen, BH 1988, 'Experiences from load tests', *Int. Symp. on Re-evaluation of Concrete Structures*, Danish Concrete Institute, Copenhagen, Denmark, pp. 153–162.

Cole, RG & Horswill, P 1988, 'Alkali-silica reaction: Val de la Mare dam, Jersey, case history', *Proc. Instn. Civ. Engrs.*, Part 1, vol. 84, pp. 1237–1259.

den Uijl, JA, Kaptijn, N & Walraven, JC 2000, 'Shear resistance of flat slab bridges affected by ASR', *11th Int. Conf. on AAR*, Quebec City, Canada, pp. 1129–1138.

Elvin, AA & Uzoegbo, HC 2010, 'Response of a full scale dry-stack masonry structure subject to experimentally applied earthquake loading', *Journal of the South African Institute of Civil Engineering*, accepted for publication.

Elvin, NG, Lajnef, N & Elvin, AA 2006, 'Feasibility of structural monitoring with vibration powered sensors', *Smart Materials and Structures*, vol. 15, pp. 977–986.

Freeme, CR & Shackel B 1981, 'Evaluation of a concrete pavement affected by the alkali-aggregate reaction using a heavy vehicle simulator', *5th Int. Conf. on Alkali-Aggregate Reaction in Concrete*, Cape Town, South Africa, Paper S252-20.

Gallias, JL 2000, 'Comparison of damaging criteria for testing aggregates by autoclaving treatment', *11th Int. Conf. on AAR in Concrete*, Quebec City, Canada, pp. 949–958.

Gräbe, PJ & Oberholster, RE 2000, 'Programme for the treatment and replacement of ASR affected concrete sleepers in the Sishen-Saldanha railway line', *11th Int. Conf. on AAR*, Quebec City, Canada, pp. 1059–1068.

Haurylkjiewicz, J 1979, 'Critical analysis of the method of safety factors in geotechnics', *Proceedings 3rd Int. Conf. on Applications of Statistics and Probability in Soil and Structural Engineering*, Sydney, Australia, vol. II, pp. 672–679.

Hendry, AW 1964, *Elements of experimental stress analysis*, Pergamon, Oxford, U.K.

Ho, K, Leroi, E & Roberds, WJ 2000, *Quantitative risk assessment: applications, myths and future direction*, GeoEng 2000, Melbourne, Australia, pp. 269–312.

Hohenbichler, M & Rackwitz, R 1988, *Improvement of second-order reliability estimates by importance sampling, Jour. of Eng. Mech.*, ASCE, vol. 114, no. 12, pp. 2195–2199.

Holm, J, Idorn, GM & Braestrup, MW 1988, 'Investigation, re-evaluation and monitoring of the Lucinda jetty prestressed concrete roadway', *Int. Symp. on Re-evaluation of Concrete Structures*, Danish Concrete Institute, Copenhagen, Denmark, pp. 119–131.

Imai, H, Yamasaki, T, Maehara, H & Miyagawa, T 1986, 'The deterioration of alkali-silica reaction of Hanshin expressway concrete structures – investigation and repair', *7th Int. Conf. on AAR in Concrete*, Ottawa, Canada, pp. 131–135.

Kojima, T, Hayashi, H, Kawamura, M & Kuzume, K 2000, 'Maintenance of highway structures affected by alkali-aggregate reaction', *11th Int. Conf. on AAR in Concrete*, Quebec City, Canada, pp. 1159–1166.

Leemann, A, Thalmann, C & Studer, W 2004, 'AAR in underground structures of Switzerland – a survey', *12th Int. Conf. on AAR in Concrete*, Beijing, China, pp. 1071–1077.

Meo, M & Zumpano, G 2005, 'On the optimal sensor placement techniques for a bridge structure', *Engineering Structures*, vol. 27, pp. 1488–1497.

National Institute for Transport and Road Research 1981, *Code of practice for the design of highway bridges and culverts in South Africa*. TMH7, Parts 1 and 2. National Institute for Transport and Road Research, Pretoria, South Africa.

Oberholster, RE, Maree, JS & Brand, JHB 1992, 'Cracked prestressed concrete railway sleepers: alkali-silica reaction or delayed ettringite formation?', *9th Int. Conf. on AAR in Concrete*, London, U.K., pp. 739–749.

Ockleston, AJ 1963, 'Test on portal beam model, Goch street viaduct', *Unpublished Report to City Engineer*, Johannesburg, South Africa.

Ono, K 1989, 'Assessment and repair of damaged concrete structures', *8th Int. Conf. on AAR in Concrete*, Kyoto, Japan, pp. 647–658.

Ono, K & Taguchi, M 2000, 'Long term behaviour of AAR bridge pier and the internal deterioration', *11th Int. Conf. on AAR in Concrete*, Quebec City, Canada, pp. 1167–1174.

Rackwitz, R & Fiessler, B 1978, 'Structural reliability under combined random load sequences', *Computers & Structures*, vol. 9, pp. 484–494.

Semelink, CJ 1981, 'Field survey of the extent of cracking and other details of concrete structures showing deterioration due to alkali-aggregate reactions in the South Western Cape

Province', *5th Int. Conf. on Alkali-Aggregate Reaction in Concrete*, Cape Town, South Africa, Paper S252/19.

Shayan, A, Al-Mahaidi, R & Xu, A 2008, 'Durability and strength assessment of AAR-affected bridge deck planks', *13th Int. Conf. on AAR in Concrete*, Trondheim, Norway, p. 11 (Available on CD, pages not numbered.).

US National Research Council 1989, *Improving risk*, Communication Committee on Risk Perception and Communication, National Academy Press, Washington, DC, U.S.A.

Udd, E 1996, 'Fibre optic smart structures', *Proceedings of IEEE*, vol. 84, no. 1.

Whitman, RV 1997, 'Acceptable risk and decision-making criteria', *International Workshop on Risk-based Dam Safety Evaluation*, Trondheim, Norway.

PLATES

Plate 4.1 A cracked coal load-out silo kept in daily operation while being investigated for possible structural inadequacy. The silo was filled and emptied completely, twice daily.

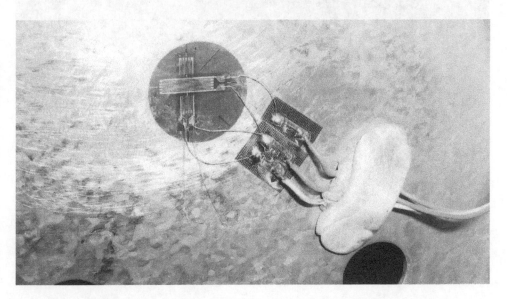

Plate 4.2 Rosette of two electric resistance strain (ERS) gauges mounted on surface of steel structural member.

Plate 4.3 Mechanical dial gauge being used to measure deflection of beam in situ under load.

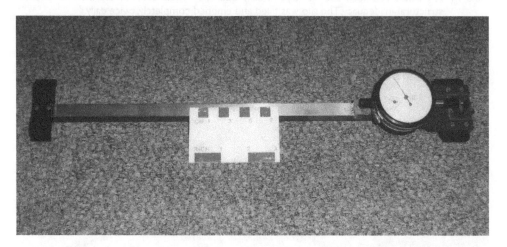

Plate 4.4 400 mm gauge length 'Demec' mechanical strain gauge incorporating a mechanical dial gauge.

Plate 4.5 LVDT (linearly variable displacement transducer). LVDTs can be used to replace dial gauges in almost all applications.

Plate 4.6 ERS gauges mounted in pairs on exposed reinforcing of a reinforced concrete column damaged by AAR.

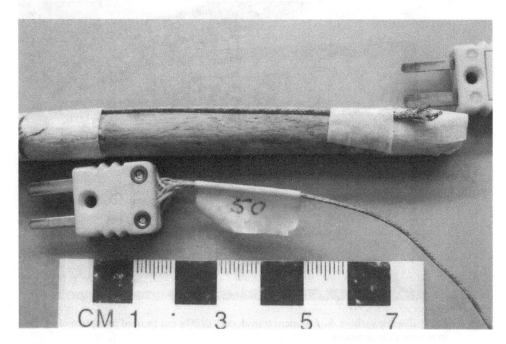

Plate 4.7 Thermocouples used to measure temperature in either field or laboratory applications. Two junctions appear in the plate, both taped to the wooden dowel.

Plate 4.8 Bubble level gauge mounted on bracket bolted to an AAR-damaged beam to measure change in slope with load.

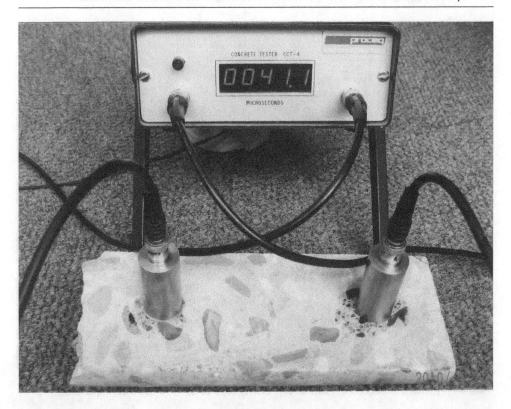

Plate 4.9 Ultra-sonic pulse velocity (UPV) gauge recording UPV transit time of 41.1 microseconds over a gauge length of 150 mm (indirect measurement mode).

Plate 4.10 Portal frame being prepared for in situ loading test (Dark spots on beam are application points for UPV measurements shown in Figure 4.9a.).

Plate 4.11 Remains of section of Hanshin expressway toppled by 1995 Kobe earthquake.

Plate 4.12 Weighed, loaded trucks being placed on the deck of an overhead highway structure for a test loading. Ultimately, all four traffic lanes were loaded.

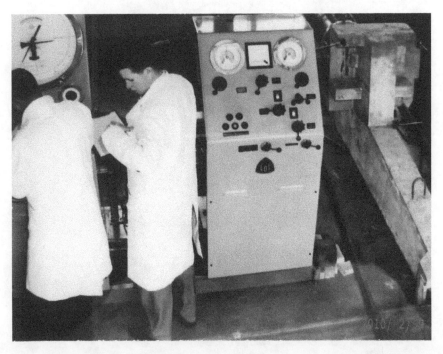

Plate 4.13 Section of a model of the portal frame shown in Plate 4.10, being tested in laboratory before constructing prototypes.

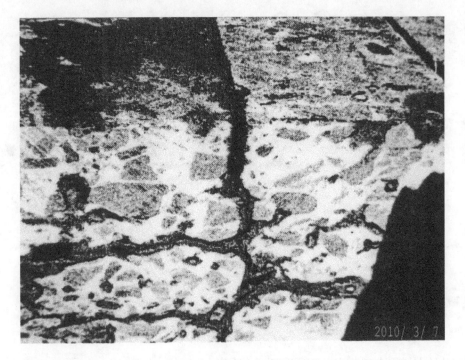

Plate 4.14 Cracks in a plane concrete highway slab damaged by AAR.

Plate 4.15 Surface cracking associated with cracks shown in Plate 4.14. (Also see Figure 4.20).

Plate 4.16 Underside of Lucinda jetty (Section 4.11.1).

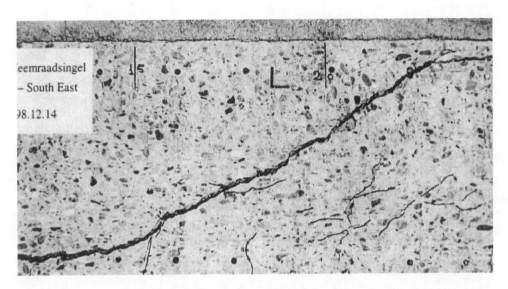

Plate 4.17 Shear crack in plain concrete deck slab removed from bridge and tested in laboratory (den Uijl, et al., 2000).

Plate 4.18 Side view of prestressed concrete plank taken from AAR-damaged structure and tested to failure load (Shayan, et al., 2008).

Chapter 5

Repair and rehabilitation of AAR-affected structures

5.1 TYPES OF REPAIR OR REMEDIAL TREATMENT

It was shown in Chapter 4 that although the cracking caused in concrete by AAR may be visually alarming, there is strong evidence that the effects of the deterioration may be less serious in terms of structural safety than appearances might suggest. This is at least partly because most structural loading codes are almost unrealistically conservative. Thus building and highway structures are hardly ever required to carry more than a small proportion of their design loads during service, and actual service stresses are generally low. The latter statement has also been demonstrated by measuring dynamic strains in a bridge pier under real service loading (Section 4.4.1). Visually the pier had deteriorated severely as a result of AAR. However, the actual service live load compressive stresses in the concrete amounted to a maximum of only 0.5 MPa. As the minimum strength measured for cores taken from the pier was 28 MPa it was most unlikely that there could be an immediate safety problem. Full-scale load tests on structures showing AAR deterioration have been carried out in many countries, where a range of climates is experienced. In some cases these tests have been repeated over a number of years. In all cases the structures were found to behave predictably with little loss in serviceability as a result of the deterioration by AAR.

If the margin of safety of a structure that has been attacked by AAR remains adequate, rehabilitation of the structure and repair of the AAR-related damage must inevitably come under consideration. The object of this chapter is to survey the quite extensive experience of repairs and repair methods that has appeared in the literature over the past 30 years.

The main difficulty in writing this Chapter has not been a scarcity of ingenious repairs and repair methods, it has been the lack of follow-up studies and reports (several years later) of the success or problems experienced with the repairs. This is common in all fields of civil engineering. Many papers are written describing ground-breaking new methods. However, whether in the course of time, these methods prove to be either satisfactory or unsatisfactory, it is very seldom that the original authors write a paper reporting and analysing the outcome and explaining why the method was either successful or unsatisfactory.

The types of repair that will be discussed and evaluated by considering successful or less successful examples will be of three general types:

1   Arresting or avoiding the AAR process, usually by surface treatment;
2   Correcting structural deficiencies that have resulted from AAR by external treatments; and
3   Radical treatment - partial or complete demolition and reconstruction.

5.2   ARRESTING THE AAR PROCESS – EXPERIMENTS WITH SURFACE TREATMENTS

In many cases, arrest of the AAR process may be considered a pre-requisite to other forms of treatment. There may appear to be little point in applying a surface treatment to cover the effects of AAR if the reaction is not rendered and kept dormant. Otherwise, cracking will recur within a few years (see, e.g., Plates 1.5 and 1.6). In many cases, arrest and dormancy of AAR may be considered the only treatment required, together with a surface treatment to improve the appearance of the structure. Unfortunately, it is the form of treatment which is least understood and which has the least satisfactory record of success.

Sixty years ago Vivian (1950) showed that the amount of expansion that occurs in cement mortars as a result of AAR depends on the amount of removable water in the mortar. Vivian's original results are reproduced in Figure 5.1. If the removable water was less than 4% by dry mass, no expansion due to AAR occurred. Once the removable water exceeded 4% the expansion became directly proportional to the excess of removable water over 4%. Vivian called this excess the available water. Removable water was defined as the water lost after prolonged storage over calcium chloride (a relative humidity of 32%). The available water is that part of the total water that is held in capillaries in the mortar. Note from Figure 5.1 that the available water may be contained within the concrete ab initio (sealed specimens) or be allowed to penetrate the mortar from outside (unsealed specimens) after some drying has occurred. Assuming that Vivian's results on mortar are applicable to concrete, it would appear that if the available water in concrete can be kept below an upper limit, expansion by AAR can be reduced or eliminated.

Other early evidence (Gudmundsson and Asgeirsson, 1983, Nilsson, 1983, Ludwig, 1989) has suggested that if the relative humidity (RH) in the atmosphere immediately surrounding a concrete structure can be maintained at below 95%, AAR will be inhibited. However, the relative humidity of the surroundings is unlikely to be the same as the relative humidity in the pores of the concrete where the AAR takes place. Water in the pores of concrete maintained at a relative humidity of less than 100% is in a state of tension or suction. The pore water tension is balanced by compressive stress in the solid components of the concrete. The net effect is that any expansive stresses are counterbalanced, partly or completely, by the pore water tension or suction, as well as by tension induced in the reinforcing, confinement by other parts of the structure, e.g. friction on the base of a road pavement or by existing applied prestress.

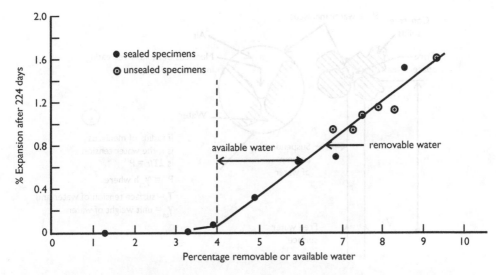

Figure 5.1 Vivian's (1950) observed relationship between removable water and expansion.

The relative humidity (RH) in a porous material can be related to the pore water suction or negative pressure p^{II} by the Kelvin equation (e.g. Blight, 1966), see Figure 5.2a:

$$p^{II} = 311000 \log_{10} (RH) = 2T/r \text{ in kPa} \tag{5.1}$$

where RH is expressed as a fraction of unity, e.g. 0.95. T is the surface tension of water in kPam, r is the radius of the water menisci in m. Because RH will be less than 1.0, the value of p^{II} will be negative, indicating that p^{II} is tensile.

At a RH of 0.95, the suction is nearly 7000 kPa. Even if the suction is not completely effective in compressing the concrete there must, at RH = 0.95, be an isotropic compressive stress in the concrete of several thousands of kPa that balances the tension in the pore water.

The principle of the balancing of a tension in the pore water by compression in the concrete solids has been demonstrated experimentally by Blight (1966, 2008). In the 2008 experiments, specimens of concrete that had been attacked by AAR were air dried at relative humidities of 30 to 40% and temperatures between 20 and 25°C. As shown in the upper part of Figure 5.2b, when exposed to RH 100%, the concrete immediately started to swell and after 60 days had swelled by a strain of 740×10^{-6}. The lower part of Figure 5.2b shows that the concrete absorbed 2.7% of water in the swelling process. The specimens were then dried to constant mass at a temperature of 50°C and after cooling, were exposed to controlled relative humidities of 86, 93 and 98% at a temperature controlled to 20° ± 1°C. The lower part of Figure 5.2b shows that the three concrete specimens (A) started absorbing water, similarly to when they had been air-dried at 20–25°C, but as the upper part of the diagram shows, all

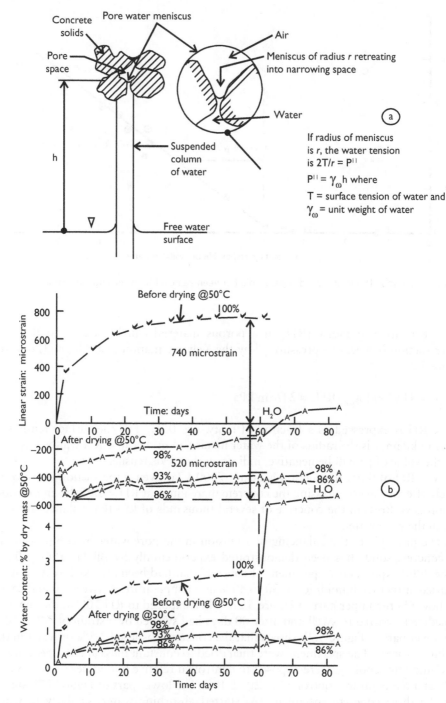

Figure 5.2 (a) Capillary model of partly saturated concrete, on which the Kelvin equation is based. (b) Behaviour of specimens "A" during exposure to relative humidities of 86% to 98% at 20°C after drying at 50°C ("⌄" shows behaviour of specimen exposed to nominal 100% RH before drying at 50°C. ("A" specimens were AAR-affected.).

specimens at first compressed, by as much as 520×10^{-6}, and then, after 2 to 5 days, started to swell. The absorption of water, however, was continual and uninterrupted, but less than before drying at 50°C.

The explanation of the initial compression is that drying at 50°C removed all of the removable water from the pores of the concrete, hence the Kelvin equation no longer applied and the compressive stress applied by tension in the removable pore water disappeared. When re-exposed to high relative humidities, water vapour condensed in the pores, re-established the menisci and the pore water tension and the concrete was re-compressed to develop stresses that balanced this tension. Once the menisci had fully re-formed (see Figure 5.2a) their radii began to increase, increasing the RH in the pores, reducing p^{II} and allowing the concrete to swell (equation 5.1).

These tests (replicated and repeated) showed that the principle expressed by equation (5.1) applies over the whole range of possible RH values. Hence one obvious way of inhibiting AAR is to dry the interior of the concrete out until the suction exceeds the swelling pressure developed by the AAR, and then maintain it in that condition by means of a waterproofing layer or coating. Based on observed values of the swelling pressure exerted by AAR, it appears that if the suction within the concrete can be maintained at above 5000 kPa (RH below 0.97) expansion caused by AAR should be eliminated or much reduced.

It appears fairly simple to achieve this aim in the laboratory, as many studies have demonstrated. Four sets of field tests will be referred to here, as field studies have shown that laboratory results often cannot be replicated on a large scale in the field.

5.2.1 Experiments in Iceland (cold climate) and France (cool temperate climate)

Some of the very early research on suppressing AAR by drying out concrete and keeping it dry, was done in Iceland (Gudmunsson and Asgeirsson, 1983, Nilsson, 1983). Olafsson (1989) describes tests on the unpainted concrete walls of a house in Iceland, the results of which are shown in Figure 5.3. The reader is not told if the surfaces of the walls were protected from impinging rain and had the same or different orientations, although driving rain is mentioned as one of the causes of damp walls. Nevertheless, it is indisputable that both walls dried out progressively for at least 3 years after being treated.

Godart, et al., (1996) undertook laboratory tests to find surface treatments for concrete that would be effective in preventing the ingress of moisture through the surface. Their $70 \times 70 \times 280$ mm prismatic test specimens were dried in a 50% RH atmosphere for 3 months before coating the surfaces according to the instructions of the various coating manufacturers. After allowing the coatings to cure for 7 days, the specimens (in addition to an untreated control and a specimen wrapped in two layers of aluminium foil that was sealed to the concrete with epoxy resin) were placed in sealed containers containing trays of water to humidify the air. The containers were stored at 38°C and the specimens were weighed and their swell elongation was measured over a 6 month period. The results of the experiment are shown in Figure 5.4. (Note that the order of increases in water content do not always agree with the order of increasing elongation. This may be partly caused by variations in the concrete from specimen to specimen.)

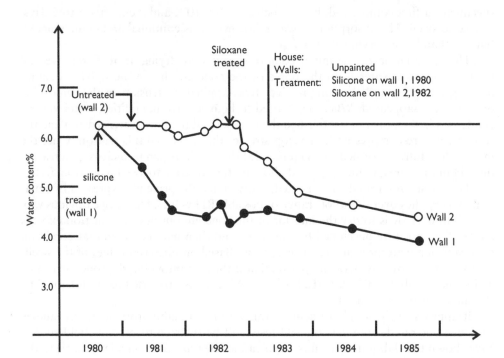

Figure 5.3 Moisture reduction in concrete house walls in Iceland after impregnation with silicone and siloxane.

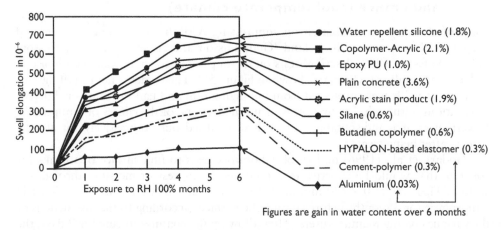

Figure 5.4 Water absorption tests on treated 70 × 70 × 280 mm concrete specimens (Godart, et al., 1996).

Based on the results of this experiment, a large scale field experiment was undertaken on a road-over-road bridge on the A4 motorway near Paris. The bridge was built in 1976 and the appearance of AAR was first noted in 1986. In 1990 it was decided to attempt to waterproof the entire surface of the concrete structure to arrest the AAR. The deck was stripped to bare concrete and a coating consisting of a cold impregnation coating, a bituminous membrane, a protective layer of asphalt and a gravel coated wearing course were applied. The entire underside of the deck and the surfaces of the piers were coated with a 3 mm thick layer of polymer-modified cement mortar (the "cement-polymer" of Figure 5.4). Direct pull-off tests showed that the mortar's adhesion strength to the cleaned concrete substrate was between 1 and 2 MPa.

Four years later, in 1994, an inspection found that cracking had reflected through the mortar layer, which had already been repaired locally. The re-opened cracks were accompanied by a white efflorescence. There was no overall longitudinal expansion of the deck, but this had probably been restrained by the longitudinal reinforcing.

The reason for the lack of success seems to have been that although the mortar may have prevented the ingress of moisture into the concrete, as in the laboratory experiment, it also sealed in the moisture that was already present, and this sealed-in moisture allowed the AAR to continue. As stated above, it is necessary to dry the concrete out to an internal RH of 97% or less, and to maintain the dry state before AAR will be arrested. Godart, et al., do not report having monitored the interior water content of the concrete at that time, either before or after the experimental waterproofing, and presumably did not do so.

As a sequel to this study a system has been set up for long-term monitoring of dimensional changes in nine repaired, water-proof coated, AAR-affected bridges on the A26 highway in north eastern France (Delaby, Brouxel and Pascal, 2004). The system uses remotely-sensed LVDTs. These have recorded a range of different forms of movement ranging from reversible seasonal movements, indicating no net AAR swelling, to small progressive expansions of 0.05 to 0.2 mm/y over periods of 4 to 5y. Hence the AAR process in these bridges appears to have stabilized at a low rate that could be accepted as a successful outcome.

5.2.2 Laboratory experiments in South Africa (warm temperate, water-deficient continental climate)

Work on suppressing AAR by the application of surface treatments was started by the South African National Building Research Institute in 1980 (Putterill and Oberholster, 1985). A series of large beams ("large" is not defined, and the specimens were presumably unreinforced as reinforcing is not mentioned) was treated with up to 19 surface treatments. Not all coatings are described, but they included acrylic PVA, magnesium fluosilicate, polyurethane and silicone. The specimens were exposed on north-facing racks outside (in the southern hemisphere, hence facing the sun) and their expansion was measured by means of Demec gauges. The expansion was either initially inhibited by the surface treatments or took some years to develop (Figure 5.5a), but eventually, either the treatments became ineffective, or the AAR reaction developed, or both. Figure 5.5b suggests, from the changes in mass and the expansion, that after about a year, the surface treatment became ineffective and the AAR expansion developed. The overall result, however, was that the surface treatments proved ineffective.

These tests were performed on the coast in a mild climate with seasonal rain. The next set of tests was performed inland at an altitude of 1700 m in a climate having summer rainfall and dry cool winters.

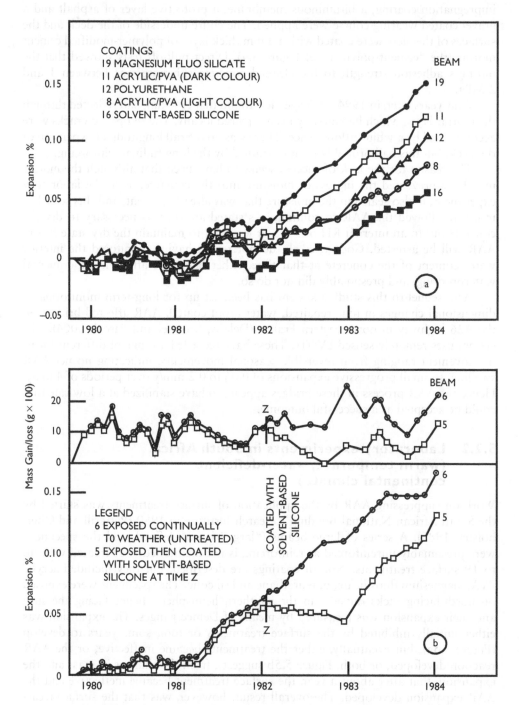

Figure 5.5 Experiments to study the effects of various surface treatments on preventing or suppressing AAR-expansion of concrete beams exposed to weather (Putterill and Oberholster, 1985).

After a survey of the literature of sealant materials for concrete, Blight (1990, 1991) concluded that the most promising surface treatments for inhibiting AAR were those that impeded water entry into a concrete structure, but also allowed free water that either was already in or that periodically seeped into the concrete, to leave by evaporation through pores in the surface treatment. In other words, the treatment should be like the human skin, impenetrable to water from the outside, but able to pass water outward from the inside.

Five types of water proofing systems for concrete were identified:

a   pore liner penetrants, usually operating as water repellents;
b   pore blocker penetrants that seal surface pores;
c   sealers that form an impervious surface skin on the concrete;
d   coatings that form a thick impervious surface skin;
e   renderings (thick coatings), usually applied by trowel.

Two pore liner penetrants and two coatings were selected for a preliminary study. The descriptions given by the suppliers were

a   silicone – a silicone in a hydrocarbon solvent;
b   silane – an alkyl alkoxy silane;
c   cement slurry – a cement-based resin-modified slurry;
d   resin emulsion – an aqueous emulsion of synthetic resins.

Laboratory tests were performed on 100 mm × 100 mm × 200 mm prisms of concrete which were oven dried to constant mass at 50°C. A set of four prisms was treated with each of the waterproofers and four were set aside as untreated controls. To avoid damage to the coatings during handling, the eight corners of each prism were each protected by a small galvanized steel angle. Each set of four angles was held in place with a band of galvanized wire. The specimens were weighed and placed on racks in a fog room where they were subjected to a continuous mist spray. Periodically, the specimens were removed from the fog, surface dried and weighed.

At the end of the test the specimens were placed on a rack on the flat roof of the laboratory building so that they were fully exposed to sun and weather. The specimens were orientated with their long axis east-west so that at least two sides had maximum sun exposure and were rotated every six months to spread the effects of weathering more evenly. The exposure was particularly severe because Johannesburg averages 340 days of sunshine per year at its elevation of 1800 m. Due to the semi-arid climate (700 mm of annual precipitation and 1600 mm evaporation) and altitude, temperature stressing and ultra-violet exposure are severe. After four years of exposure, the original fog room test was repeated.

Figure 5.6 shows the results of the measurements. Each data point is the mean of weighings of all four specimens for a particular treatment. None of the treatments actually waterproofed the concrete, but the penetrants were considerably more successful at inhibiting water entry than the coatings. The control specimens initially absorbed 5.5% of moisture, and the coated specimens eventually absorbed almost as much. The silane limited moisture absorption to 1%, but the silicone merely retarded the rate of absorption and did not ultimately limit it.

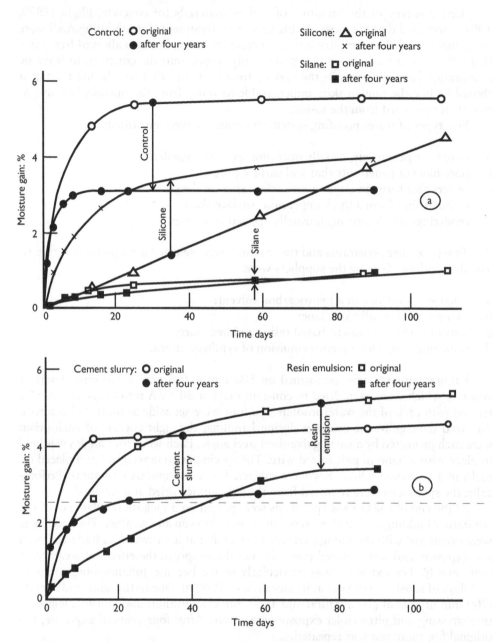

Figure 5.6 (a) Variation of moisture gain with time for control specimens and specimens treated with silicone and silane. (b) Variation of moisture gain with time for specimens treated with cement slurry and resin emulsion.

After four years of exposure, all the specimens absorbed less moisture, including the control specimens. Absorption by the control specimens and the coated specimens had decreased by about 45%. The silane was virtually unaffected by 4 years' exposure, while the silicone (as compared to the control) no longer had any effect.

This agrees with the observations recorded in Figure 5.5b. Water absorption of the silicone-treated specimens was now actually greater than that of the control specimens, but still substantially less than the initial water absorption of the controls. Hence, only the silane emerged from this series of tests with any credibility as a waterproofer.

The most obvious cause of the decrease of water absorption appeared to be carbonation of the concrete during its period of exposure. The depth of carbonation (revealed by spraying broken surfaces with phenolphthalein) proved to be 8–10 mm, except for the silane-treated specimens, where the depth of carbonation was only 5 mm. The coated and control specimens showed a similar depth of carbonation to that of the silicone-treated concrete.

Oxygen permeability tests on slices of concrete cut parallel to the outer faces of the prisms, and with flow directed normal to their exterior surfaces, showed that the permeability of untreated uncarbonated concrete was about five times that of the carbonated exterior surfaces of the control specimens, and that the oxygen permeability of the silane-treated surfaces was about twice that of the other surfaces. The silane was thus, after four years, still acting as a water repellent, as well as inhibiting carbonation. On the assumption that water permeability is proportional to oxygen permeability, it is postulated that carbonation resulted in reduced water absorption.

In a series of laboratory tests which simulated the effects of the South African climate, specimens similar to the 4 year old exposed specimens were prepared by soaking in water for seven days. They were then allowed to dry for one day, after which the surface treatments were applied. The specimens were subjected to a cycle starting with seven days in the fog room followed by seven days over a saturated calcium chloride solution which controlled the humidity to 0.32 (32%). Thereafter, to simulate a climate of short wet spells followed by much longer dry spells, the cycle comprised 6 h in the fog followed by seven days at RH = 0.32. The results of this treatment are shown in Figure 5.7. It can be seen that all of the specimens, including the untreated control, dried out progressively. This lead to a tentative conclusion that in a semi-arid climate, either

a   it is not necessary to waterproof exposed uncracked concrete because it will dry out completely between brief wet spells, or

b   the waterproofing of uncracked concrete need not be 100% effective: provided that most of the incident water is excluded, the concrete will subsequently dry out.

5.2.3   Field experiments in South Africa

A series of tests on full-size structures in the field was undertaken to test these tentative conclusions (Blight, 1990). A group of columns supporting overhead motorway structures was selected for these tests: one of the five test columns is shown in Plate 5.1. The columns were chosen because they are completely protected from incident rainfall by the structures overhead. Thus the moisture condition of the column surfaces could be controlled even though they are outside structures. However, it will be noted from Plate 5.1 that some water does run down the column mushrooms from above. Vertical cracks caused by AAR are also visible in Plate 5.1. The columns each contain an axial drainage duct provided to drain storm water from the roadway above. Because blockages were experienced, the inlets to these ducts had been sealed

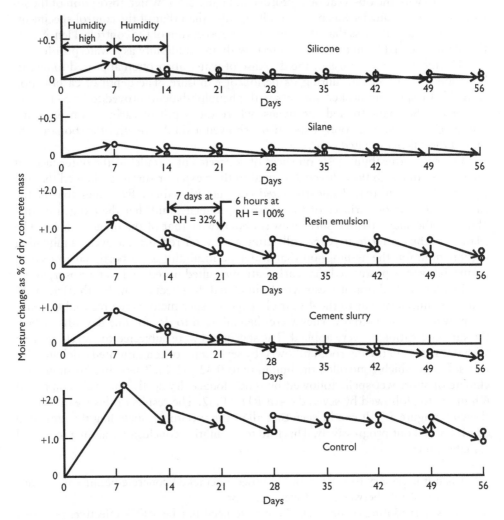

Figure 5.7 Progressive drying (measured as percentage moisture gain) of treated and untreated concrete specimens in alternating low and high humidity environments: low RH = 32%; high
RH = 100%.

some years previously and the drainage was redirected by way of external metal ducts.
The outlets of the internal ducts were left open. (The external storm water drainage
pipe can be seen in Plate 5.1).

The column shafts were cleaned with high-pressure water jets and after leaving
to dry for 7 days, each was treated with one of the four waterproofers (listed in Section 5.2.2) under the supervision of the suppliers. A fifth column was cleaned and
reserved as a control.

The moisture condition within the concrete was assessed by inserting thermocouple psychrometer probes to a depth of 100 mm into holes drilled into the concrete. A psychrometer (see, e.g. Savage and Cass, 1984) consists of a miniature

thermocouple sealed into a small cavity in the concrete. Plate 5.2 shows the tip of a commercially available psychrometer. The miniature thermocouple junction is inside the protective stainless steel wire mesh cap. The air surrounding the thermocouple comes to temperature and moisture equilibrium with the adjacent concrete. A current is passed through the thermocouple, the junction of which cools by the Peltier effect until its temperature falls below the dew point for the moisture in the surrounding air. A minute droplet of condensation then forms on the cooled thermocouple junction. The cooling current is interrupted and the condensed droplet starts to re-evaporate at the dew point temperature. The thermocouple is used to measure this temperature and hence the RH in the cavity at the end of the hole can be determined. The measuring range of a psychrometer is from RH = 1.00 to RH 0.96. Below RH = 0.96 it is difficult to condense water out of the atmosphere by cooling the air.

As recorded earlier, the relative humidity in the pores of a porous material can be related to the tension or suction p^{II} in the pore water by means of the Kelvin equation (see equation 5.1).

Even at RH = 0.96, the suction is 5500 kPa. This can be thought of as the tensile stress with which the concrete draws in any water available on its surface. Hence, a low p^{II} corresponds to moist concrete and a high p^{II} corresponds to dry concrete.

It should be noted that p^{II} has two components, a matrix component that arises from capillary tension forces in the concrete and an osmotic component arising from the dissolved salts in the pore water. The osmotic component in concrete can be as high as 2000 kPa, even when the concrete is wet.

More recently, Bakker (2004) has reported using relative humidity and temperature sensors to monitor the moisture condition in reinforced concrete structures. These appear also to be thermocouple psychrometers. Bakker has also used TDR (Time Domain Reflectometry) sensors that monitor the moisture condition in concrete via its electrical impedance. There are also sensors available that sense moisture via the changing electrical capacitance of the concrete. In the authors' experience, however, both of these methods are difficult to calibrate for use in AAR-affected concrete. They have to be calibrated under laboratory conditions and there is no guarantee that the concrete in the structure being monitored will have identical impedance or capacitance properties versus moisture content properties. Most importantly, the presence or development of cracks will drastically alter both the impedance and capacitance and AAR-affected concrete will, by definition, contain an abundance of cracks. Hence not only will the calibration probably change from the laboratory to the field, but, if cracking is occurring, the calibration will change progressively, with time.

Jensen (2004) has promoted the use of the absorbance of moisture by wooden dowel sticks. This is a very simple and basic method of monitoring moisture, but the problem again lies in the difficulties of calibration between the water content of a wooden stick and the relative humidity of the concrete. Only a thermocouple psychrometer is capable of measuring the relative humidity in the pores of concrete completely independently.

Figure 5.8a shows some preliminary measurements on the control and the silicone-treated columns. Initial readings were taken at the end of the dry season during a period of hot sunny weather. The suction in the control column was almost at the limit of measurement (5500 kPa), but that in the silicone-treated column was much less (2000 kPa). A spell of heavy rain followed, during which the suction in the

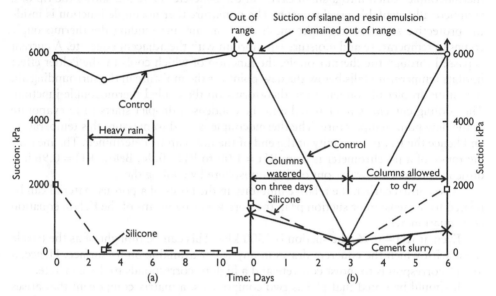

Figure 5.8 Results of first measurements on columns: (a) preliminary tests; (b) watering tests.

control column decreased slightly while that in the silicone-treated column almost disappeared. Shortly afterwards, streaks of what was assumed to be silica gel from the AAR exuded from cracks in the silicone treated column and mushroom head. Similar streaks had appeared previously and were visible on some of the other columns. It appeared that water was entering the concrete from above, probably by way of the imperfectly sealed central drainage duct. When the rain was succeeded by sunny weather the suction in the control column increased and went out of range, while that in the silicone-treated column remained depressed.

A few weeks later, during a spell of hot dry weather, the suction in each of the columns was measured. It was found that p^{II} in the control, silane- and resin emulsion-treated columns was out of range. In the silicone- and cement slurry-treated columns, p^{II} was ~ 1000–1500 kPa. The exterior surfaces of the columns were then sprayed with water and kept wet for 8 hours per day on three consecutive days to simulate a spell of natural wet weather (Figure 5.8(b)). The suction in the silane- and resin emulsion-treated columns remained out of range, while that in the control fell to 350 kPa. The suction in the silicone- and cement slurry-treated columns also responded by falling to low values. Four days later, the suction in the control column was again out of range, while that in the silicone- and cement slurry-treated columns had partially recovered.

These tests confirmed the earlier conclusions that if there is no internal source of water, sound concrete (e.g. the control column) will quickly dry out after being surface wetted. Sound, uncracked concrete probably does not need to be waterproofed in semi-arid climatic conditions.

Four years later the suction in the five columns was measured over a period of 28 days during a hot dry spell (Figure 5.9). The suction in all of the treated columns

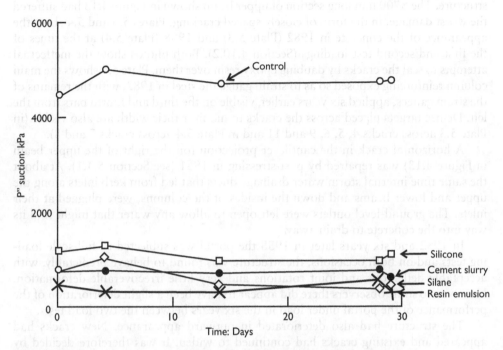

Figure 5.9 Results of measurements on columns four years after waterproofing.

was low, and remained depressed throughout the period of measurement. The suction in the control column was at its level of four years previously. These results were so surprising that at first the psychrometers were thought to be defective. They were all changed, with replacements being tested before installation, but the measurements remained the same. When the final reading was made on the silicone-treated column, water actually ran out of the psychrometer hole. Although the suctions recorded in Figure 5.9 are not zero, it is believed that the matrix suction is very close to zero and that the low values in the figure mainly represent osmotic suction.

In the light of the earlier observations on the silicone-treated column, and the exudation of gel from some of the other columns, the only possible conclusion seems to be that defective seals at the top of the internal drainage ducts were allowing small quantities of water to enter the concrete. Before the surfaces of the columns were treated, this moisture could be passed out through the concrete surface by evaporation, as presumably still happens with the control column. The waterproofing treatments have, however, retarded this moisture loss sufficiently for water to accumulate within the concrete, thus permanently depressing suctions. For concrete susceptible to AAR, this could be disastrous, as the AAR would be promoted instead of being arrested. Hence waterproofing a structure that is fed water from an internal source could be the worst measure to undertake.

The portal frame shown in Figure 4.13 became the subject of intense investigation, firstly to establish its structural safety (Section 4.9.2) then to investigate means of halting the reaction (this section) and finally (Section 5.6.3) to repair and rehabilitate the structure. The 8700 mm long section of upper beam shown in Figure 4.13 had suffered the worst damage, in the form of closely spaced cracking. Plates 5.3 and 5.4 show the appearance of the concrete in 1982 (Plate 5.3) and 1988 (Plate 5.4) at the times of the first and second test loadings (Section 4.10.2). Both photos show the ineffectual attempts to seal the cracks by daubing epoxy resin over them. Plate 5.4 shows the main column reinforcing exposed so as to strain gauge the steel in 1982, with the remains of the strain gauges, applied six years earlier, visible on the third and fourth bars from the left. Demec targets placed across the cracks to monitor their width are also visible (in Plate 5.3 across cracks 4, 5, 6, 9 and 11 and in Plate 5.4 across cracks 7 and 8).

A horizontal crack in the cantilever projection (on the right of the upper beam in Figure 4.13) was repaired by post-stressing in 1981 (see Section 5.4.1). At about the same time internal storm water drainage ducts that led from kerb inlets along the upper and lower beams and down the insides of the columns, were plugged at their inlets. The ground-level outlets were left open to allow any water that might find its way into the concrete to drain away.

In 1982 and six years later in 1988 the portal was subjected to full-scale loading tests and on both occasions, the structure was found to behave predictably, with acceptable deflexions and joint rotations and very little irrecoverable deformation. However, to some observers there did appear to have been a slight deterioration of the performance of the portal under load in the six years between the two load tests.

The structure had also deteriorated in outward appearance. New cracks had appeared and existing cracks had continued to widen. It was therefore decided by the owners that the structure must be rehabilitated, and investigations were started to find the most effective form of rehabilitation.

A number of possible solutions to the repair and rehabilitation of the apparently badly cracked section of beam were considered. These included:

a   filling the major surface cracks with an elastomeric sealer and sealing the surface of the exposed concrete to exclude moisture from the surface and from minor surface cracks;
b   encasing the beam in a ventilated metal sheath to exclude incident rain, but allow the concrete to dry out gradually to the surrounding atmosphere;
c   demolishing the damaged length of the upper beam and reconstructing it in reinforced concrete; and
d   various variations of (c) above, including augmenting the strength of the upper beam with bolt-on steel members, and replacing the damaged length with a bolt-on steel beam.

If measure (a) was to be adopted it would be important to seal the cracks and concrete surface at a time of the year when the concrete was at its driest. Johannesburg has well-defined wet and dry seasons, and the obvious time appeared to be August/September, at the end of the dry season. However, it was not known to what extent the moisture in the concrete varied seasonally, or if the concrete dried out to any significant extent during the dry season. To provide this information, the series of thermocouple psychrometer measurements described below was undertaken.

The width of the portal beam and columns is 1250 mm. To ensure that information would be available for most of the thickness of the concrete, the psychrometers were installed in pairs in holes drilled to depths of 50 and 400 mm in from the concrete surface. The probes were inserted into the holes at the end of wooden dowel sticks that had been heavily varnished with polyurethane varnish. The varnish prevented absorption of water by the dowels, which were push fits in the holes. Soft rubber discs attached to the ends of the dowels, through which the psychrometer leads were passed, further helped to seal off each cavity at its end. As the psychrometers were installed at a height of 16 m above ground level they had to be installed and accessed for reading via a truck-mounted hydraulic boom designed for repairing over head power lines. Plate 5.5 shows the hydraulic boom extended to reach the psychrometers installed in the portal and Plate 5.6 shows in situ measurements on the psychrometers being taken from the basket at the top of the boom.

Measurements were taken at intervals for a period of 20 months and the results have been summarized in Figures 5.10 (period from April to December 1989) and 5.11 (period from January to November 1990). Readings were discontinued after it was decided, on the basis of the measurements, to rehabilitate the portal by demolishing and reconstructing the upper portal beam.

The upper portion of each of Figures 5.10 and 5.11 shows the variation of moisture suction with time, while the lower portion shows the corresponding rainfall,

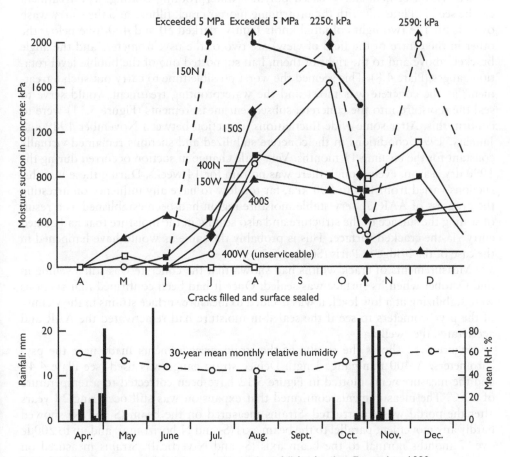

Figure 5.10 Measurements of moisture suction and rainfall for April to December, 1989.

plotted on a daily basis, and the 30-year mean monthly atmospheric relative humidity (plotted for only one year). The measurements are identified by the depth of installation and the direction faced by the side of the portal, e.g. 150 N is installed at 150 mm depth on the north-facing side of the beam. The measurements were started in April 1989, towards the end of the wet season, and it was found that suctions were negligible, indicating that the concrete was extremely wet. However, it is possible that at this stage the psychrometers were not yet in moisture equilibrium with the concrete.

During the 1989 dry season the 150 mm-deep psychrometers recorded large suctions, with 150 N (which faces the sun in winter) going out of the range of measurement and 150 S (permanently in shade in winter) recording 2250 kPa. 150 W showed a disappointingly low maximum suction of 1600 kPa. Only two of the 400 mm-deep psychrometers were in working order, and they too showed very disappointing maximum suctions of less than 1000 kPa. On the evidence presented in Figure 5.10, concrete that has cracked as a result of AAR will not dry out to below the limiting relative humidity of 97% in a single dry season. Once the 1989 dry season was ended by rain in October, suctions plummeted once again to low values.

At the end of October (against the advice of the senior author) the major surface cracks were caulked with a stiff cement mortar and all surfaces of the beam were treated with a cement-slurry based polymer waterproofing coating. The treatment can be seen in Plate 5.7 with the protruding mortar crack fillings and the slurry wash over them. The two light coloured knots faintly marked 50 and 400 (one below the other in the centre of the field of view) are two of the psychrometers and the angle bracket, above and to the right of them, had supported one of the bubble level rotation gauges (Plate 4.8). This seemed the worst possible time to carry out such a treatment, as the concrete was moist and the waterproofing treatment would serve to seal the moisture into the concrete; subsequent measurements (Figure 5.11) were to confirm this. After some wide fluctuations in suction between November 1989 and January, 1990, conditions in the concrete stabilized and suctions remained virtually constant for the ensuing 11 months. Very little change in suction occurred during the 1990 dry season, even though there was no rain for 14 weeks. During these months, suctions varied from 100 to 900 kPa, far too low to have any influence on arresting the progress of AAR. A very stable moisture region had been established as a result of sealing the surface of the structure and also sealing in the moisture that had gained entry via the cracked surface. This is probably the same as would have happened in the case of the bridge in Paris (Section 5.2.1).

Measurements of crack widths had shown that the concrete was still swelling in mid-October when its surface was sealed. Once it had been confirmed that suctions were stabilizing at a low level, it was decided to measure surface strains in the vicinity of the psychrometers to see if the sealed-in moisture had re-activated the AAR and accelerated the swelling.

Figure 5.12 shows the results of the strain measurements made near the psychrometers. A 400 mm gauge length Demec strain gauge was used (see Plate 4.4), and the measurements plotted in Figure 5.12 have been corrected to a temperature of 20°C. The measurements confirmed that expansion was still occurring 28 years after the portal was constructed. Strains measured on the beam (S and N) showed hardly any movement parallel to the beam axis (S- and N-horizontal) and up to 200µε over 7 months normal to the beam axis (S- and N-vertical). Strains measured on

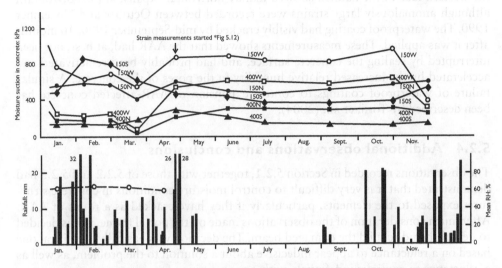

Figure 5.11 Measurements of moisture suction and rainfall for January to November 1990.

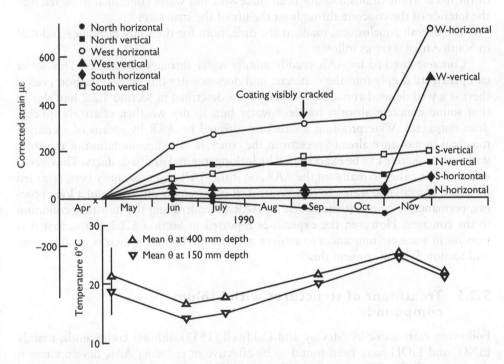

Figure 5.12 Measurements of surface strain from April 1990–November 1990.

the western face of the haunch showed mainly horizontal expansion (W-horizontal), although anomalously large strains were recorded between October and November 1990. The waterproof coating had visibly cracked by mid-September 1990, 10 months after it was applied. These measurements showed that the AAR had, at best, not been interrupted by sealing the concrete surface, and had probably been re-activated and accelerated by the increased relative humidity in the pores of the concrete. A similar failure of waterproof coatings to control expansion of AAR-affected concrete has been described by Torii, et al., (2004).

5.2.4   Additional observations and conclusions

The observations recorded in Section 5.2.1, together with those of 5.2.2 and 5.2.3 had demonstrated that it is very difficult to control moisture conditions in concrete structures exposed to the elements, particularly if they have cracked as a result of AAR. After much consideration of the observations made on the portal frame, it was decided to demolish and rebuild the damaged beam. The decision was basically a political one based on a reluctance to appear indecisive about a solution to the problem, as well as a reluctance to continue indefinitely with a programme of monitoring.

When the beam was demolished it was found that the concrete around the internal drainage ducts was visibly wet. Clearly what was observed on the columns in Section 5.2.3. was similar to the observations on the portal frame. In both cases, plugging of the inlets to the drains had not been successful and water continued to be fed into the interior of the concrete throughout the life of the structures.

The overall conclusions, made at the time, both for the study in France and that in South Africa were as follows:

Concrete cracked by AAR readily admits water through the cracks. This water can penetrate deeply into the concrete, and does not dry out by evaporation even if there is a well-defined annual dry season. Tests described in Section 5.2.2 had shown that sound concrete absorbs incident water but, in dry weather, relatively quickly dries out again. Waterproofing a structure affected by AAR by means of a coating may seal in moisture already present in the concrete. The accumulation of moisture within the concrete can be exacerbated by leaking internal drainage ducts. The overall effect will be either to maintain the AAR, or, if the AAR had previously been retarded by a lack of available water, to re-activate it. It is extremely difficult, and a long process, permanently to arrest the course of AAR by controlling the moisture condition in the concrete. However, the experience reported in Section 5.2.2 shows that it is possible in some circumstances to achieve moderate to complete success via this route and Section 5.3 will confirm this.

5.2.5   Treatment of structures with lithium compounds

Following early work by McCoy and Caldwell (1951), lithium compounds, mainly $LiNO_3$ and LiOH have been found to be effective in reducing AAR development if they are applied by spraying, to a structure suffering from AAR. According to anecdotal evidence the process is most effective in relatively dry environments, such as semi-arid climatic areas, for example New Mexico and Nevada in the United States of America. In the USA, pile caps of a bridge in South Dakota, suffering from AAR, were reportedly repaired in 1998 with concrete containing $LiNO_3$ solution. In 2004 there were no signs of further AAR in the repaired pile caps. (Larbi, et al., 2004). However, more recent research by Hooper, et al. (2004), Modry (2004) and Yin and Wen (2004) has warned that lithium compounds may cause side effects in concrete that need to be explored before lithium treatment is widely adopted in practice. The method therefore cannot yet be regarded as tried and trustworthy.

5.3 RESTORING DESIGN PROPERTIES BY RESIN-INJECTION

This section is really a continuation of 5.2, but the emphasis has shifted from surface coatings to crack injection and from swelling movements and strains to strength and elastic properties of AAR-affected concrete.

5.3.1 General consideration of crack injection as a method of repair

Portland cement grout would probably be an ideal material for injection grouting of AAR-cracked concrete were it not for two factors:

1 Injection with portland cement would locally increase the alkali content available to the concrete and possibly accelerate or reactivate the AAR; and
2 The particles of portland cement powder (even rapid-hardening cement) are of too large a size to penetrate very fine cracks.

Hence it is necessary to consider resin grouting as a medium for repairs to AAR-damaged concrete. There are a number of requirements that a resin used for injecting AAR-damaged concrete should meet:

1 It should have a sufficiently low viscosity to enable it to penetrate fine hair cracks using very moderate injection pressures.
2 It should nevertheless not have such a low viscosity that it becomes absorbed into the pores of the concrete on either side of a crack so that the crack is emptied by absorption when injection ceases.
3 Cracks generated by AAR can be expected to contain powdery or gel-like reaction products. The resin should have excellent wetting properties so that it can penetrate these products and wet the intact concrete forming the crack sides.
4 Cured resins generally have coefficients of thermal expansion that are as much as 50 times that of concrete. In the exposed field situation, cyclic thermal stresses will therefore be set up at the resin-concrete interface. Coefficients of thermal expansion for resins are in the range from 40 to $70 \times 10^{-6}/°C$ whereas for concrete the coefficient of thermal expansion is only 10 to $12 \times 10^{-6}/°C$. The resin must be able to accommodate this movement by creep, otherwise it will debond from the concrete (e.g. Blight and Mitchell, 1980). The Shaw test (Reader and Shaw, 1973) in which specimens of the resin are applied to sheets of plate glass and

thermally cycled to test their accommodation of differential thermal movement, is a useful procedure to investigate the durability of a resin-to-concrete interface, subjected to temperature fluctuations.

5   If the injection is to improve the mechanical properties of the concrete, its shear and elastic moduli should equal or exceed those of the intact concrete.

Unfortunately requirements 4 and 5 are probably incompatible.

The difficulties involved are illustrated by experience with a large (5 m × 5 m × 2.5 m deep) reinforced concrete foundation block in Johannesburg which showed severe cracking as a result of AAR. The block supports a bridge pier and is founded on undermined ground. The block was designed to be jacked in three principal directions (E-W, N-S and Up) to correct for any movement that might be caused by closure of the steeply inclined mining void at 30 m beneath it. To allow the jacking to take place, the block was housed in a reinforced concrete box that was 7 m × 7 m × 3.5 m deep in inside measurements. It was decided to attempt a repair by means of epoxy resin injection grouting. Cores of the affected concrete were taken prior to injection and further cores were taken after injection to examine the success of the operation. The resin was carefully formulated to comply with requirements 1–4 above with 4 being tested by means of the Shaw test (Reader and Shaw, 1973).

The results of the post-injection examination were disappointing. Penetration of the cracks was poor because the low allowable injection pressures were not enough to force the resin into the fine cracks. Many wide cracks also remained unpenetrated because they were not interconnected with cracks along which resin was able to flow. The resin had in many places not wetted the sides of the cracks and had not bonded to the concrete.

Figure 5.13 shows the results of a compression test on a core of epoxy-injected concrete from the foundation block. The test is compared with the results of two tests made on cores taken before the injection.

The results were again disappointing. The injected core (which was loaded and unloaded twice) was divided in two by a wide diagonal resin-filled crack. The resin had obviously imparted a reasonable strength to the core, but the compression modulus (5 GPa) and the residual deformation were clearly not acceptable. However, later experience (Section 5.3.2) has shown that very acceptable results can be obtained, using the correct techniques.

Based on the poor results of the trial resin injection, and because no movement of the block had been recorded in the past 20 years, it was decided that if any corrective jacking of the bridge was required, it could with minor modifications, be done from the top of the bridge pier. The effects of the AAR were taken care of by filling the space between the block and the surrounding RC box with concrete, thus providing confinement to the sides and base of the block.

5.3.2   Repair of sports stadium

(The data for this case history were collected by Mr Piet Coetzee and Mr Philip Ronne of BKS (Pty.) Ltd., South Africa, advised by Dr Bertie Oberholster, then of the South African Council for Scientific and Industrial Research. BKS (Pty.) Ltd. have kindly

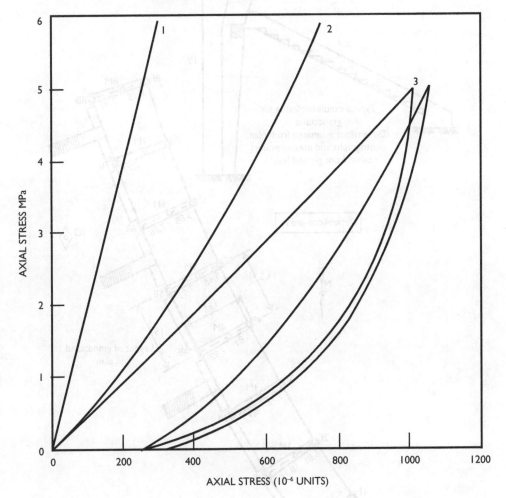

1 and 2: AAR Deteriorated concrete before epoxy injection
3 Epoxy injected AAR-Deteriorated concrete
Secant to 5MPa Moduli 1: 20.8 GPa
2: 7.6 GPa
3: 5. 0 GPa

Figure 5.13 Comparison of stress-strain curves for AAR-damaged concrete and AAR-damaged concrete that has been injected with epoxy resin.

permitted the use of the raw data by the authors who accept full responsibility for their analysis of the data and the conclusions drawn from them. We also thank the University of Stellenbosch, owners of the stadium, for permission to publish this case history.)

The reinforced and prestressed concrete stadium was completed in 1978 and Figure 5.14 shows the main dimensions in plan and a side elevation of a typical

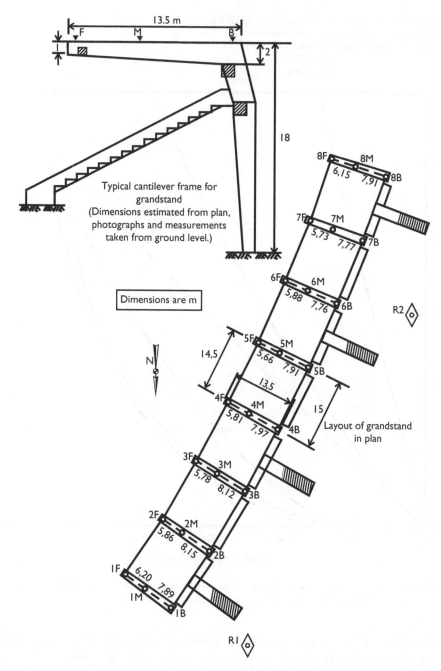

13.5 m

F   M   B

2

18

Typical cantilever frame for
grandstand
(Dimensions estimated from plan,
photographs and measurements
taken from ground level.)

Dimensions are m

8F   8M
6,15   7,91   8B

7F   7M
5,73   7,77   7B

6F   6M
5,88   7,76   6B

R2

N

5F   5M
14,5   5,66   7,91   5B

4F   13,5
4M
5,81   7,97   4B   15

Layout of grandstand
in plan

3F   3M
5,78   8,12   3B

2F   2M
5,86   8,15   2B

1F   6,20   7,89
1M   1B

R1

Figure 5.14 Plan and typical side elevation of cantilever portal frame. R1 and R2 are benchmarks.

cantilever frame. The main cantilever beam, supporting a light galvanized steel roof, was prestressed back to the column. Plate 5.8 shows an end elevation of the stadium photographed in 2006. Cracking, mainly parallel to the lengths of the beams and heights of the columns became apparent in 1987 and a full condition survey was carried out in 1989 to assess the severity of the cracking. At this time, some of the cracks were 3.5 mm wide and penetrated the 60 mm cover to the reinforcing (see Plate 5.9 which shows a wide crack running the full length of one of the cantilever beams.). It was decided to repair the structure by applying a hydrophobic surface treatment and injecting the cracks with an epoxy resin. The repair work was carried out in July/ August 1992 which was, incidentally, the end of the wet season. Plate 5.10 is a view of one of the cantilever beams showing the closely spaced injection nipples inserted along the lengths of the cracks, and also showing the light steel roofing spanning between the beams. The hydrophobic surface treatment and resin injection were followed by applying a water-proofing elastomeric coating in 1994.

Starting in 1992, precise survey measurements were made of the deflections of the beams relative to the tops of the columns at positions F, M and B, marked on the plan in Figure 5.14. Figure 5.15 shows the variation with time of the measured deflections of positions M and F relative to position B for the 10 years from July 1992 to July 2002. It will be noted that each beam from No.1 to No. 8 had deflected by a different amount relative to position B. It is also evident that from 1992 onwards, the deflection of every beam was decreasing. These upward movements have been plotted in Figure 5.16 for April 1996 and July 2002, when decreases at position F of from 3.5 to 14 mm were recorded. The most likely explanation for the continuing upward movement is that, during the period of observation, the concrete continued swelling against the prestress forces, thus pushing the beams upwards. A second possibility is that, simultaneously, the compression modulus of the concrete in the upper part of the beam section has been slowly deteriorating, allowing the prestress force to compress the top of the beam and add to the upward movement. Whatever the explanation, the net effect is on the side of safety, has occurred very slowly and Figure 5.16 shows it (in 2002) to have been occurring at an approximately constant rate.

Figure 5.15 shows the deflected shapes of the eight beams in 1992 (1^1, 2^1, etc.) and in 2002 (1, 2, etc.). It is interesting that the beam with the greatest deflection (8) has also recovered the most (Figure 5.16), with beam 3 which initially deflected least, showing the second largest recovery of deflection.

The calculated deflection profiles shown in Figure 5.15 have been calculated under the estimated dead load, using the second moment of area of the gross beam cross-section. Fitting the calculated line to the measured deflections gives a concrete elastic modulus E of 20 GPa for beams 2 and 3, 9 GPa for beam 4, 5 GPa for beams 1, 5 and 7 and 3 GPa for beam 8. All of these are very possible values: ranging from 20 GPa for almost undamaged concrete to 3 GPa for concrete severely damaged by AAR.

At the present time (2010) the structure shows no signs of new or re-opened old cracks, although this may be due to recent (unrecorded) crack filling and the application of an elastomeric paint. Plate 5.11 shows one of the columns as it appeared in 2006. The near vertical lines of the filled cracks in the column are faintly visible, but there are no recent cracks. Both visually and in terms of the beam deflection measurements, the 1992 repairs appear to have been completely successful. This case is a powerful reason for believing that, in many cases, even if AAR swelling cannot

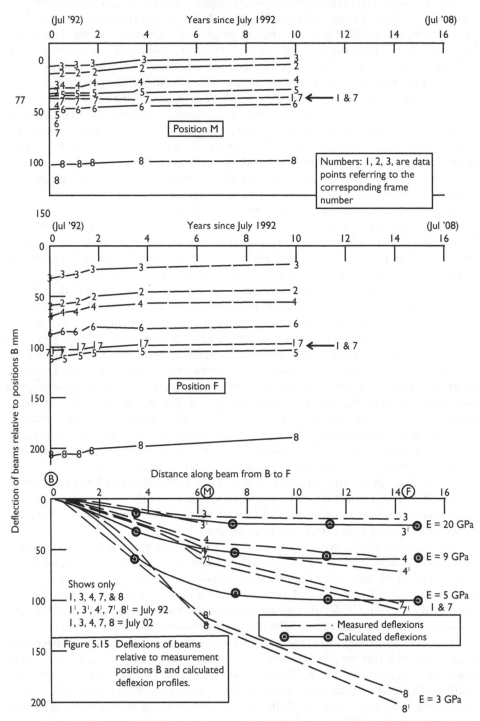

Figure 5.15 Deflexions of beams relative to measurement positions B and calculated deflexion profiles.

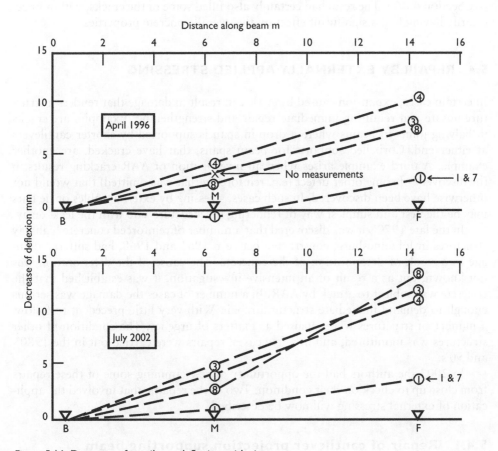

Figure 5.16 Decrease of cantilever deflexion with time.

easily be stopped, it need not always be a continuing threat to the integrity of the structure.

Other partial successes have been achieved by means of resin injection, and the Hanshin expressway described in Section 4.10.1 provides an example. Imai et al. (1986) used epoxy resin injection grouting to restore the integrity of a double cantilever beam supporting an expressway that had deteriorated by AAR. Before the repair, ultrasonic pulse velocity (UPV) measurements made on the concrete averaged 2.7 km/s (although it is not clear if the measurements were made in situ or on cores). After grouting, Imai et al report that the UPV had been restored to above 3 km/s and on this basis regarded the repair as having been successful.

Without questioning the validity of their measurements, it must be pointed out that while the increase in UPV certainly showed that the sonic discontinuities in the concrete had been reduced, the mechanical properties may not have been much improved, as a similar result would probably have ensued if the water content of the concrete had been higher when the "after injection" UPV measurements were made (see Section 4.8.7). The resin had certainly also filled some of the cracks, without necessarily having had a significant effect on the overall concrete properties.

5.4 REPAIR BY EXTERNALLY APPLIED STRESSING

In certain cases, expansion caused by AAR can result in damage that renders a structure unsafe and requiring immediate repair and strengthening. Examples are cracks in halving-joints in beams where a drop-in span is supported by shorter cantilevers at either end. Corbelled supports for beam spans, that have cracked, are another example. A third example arises when an investigation of AAR cracking results in the discovery of some other defect (e.g. reinforcing that was omitted) that would not otherwise have been discovered. In such cases, stressing by external cables or tie bars may be the best and simplest way of rendering the structure safe with the least delay.

In the late 1970's it was discovered that a number of reinforced concrete highway structures in Johannesburg, constructed between 1961 and 1968, had suffered damage as a result of deterioration of the concrete. The cause of the deterioration was not known, but as a result of an intensive investigation, it was established that the concrete was subject to attack by AAR. In a number of cases the damage was serious enough to demand immediate structural repair. With very little precedent to follow, a number of structures were repaired as matters of urgency. The condition of other structures was monitored, and in a few cases, repairs were carried out in the 1980's and 90's.

In 2003 the authors had the opportunity of re-examining some of these repairs from close up to check on their condition. Two of the repairs, that involved the application of external stressing will now be described.

5.4.1 Repair of cantilever projection supporting beam spans on either side

The cantilever projection on the east side of the portal frame shown in Figure 4.13 had, as a result of AAR, developed a horizontal crack. The position of the cantilever and the location of the crack are shown in Figure 5.17. The cantilever supports a series of simply supported beams, via halving joints, as shown in elevation in Figure 5.17, and a repair therefore had to be effected without delay. As an interim safety measure, the traffic lane adjacent to the cracked cantilever (which happened to be the slow lane) was closed to traffic.

In 1983, the repair was made by stressing the upper and lower portions of the cantilever together with one pair of external and one pair of internal high tensile steel stressing bars. Electric resistance strain gauges were glued to the concrete and the stressing bars and used to monitor strains in both steel and concrete in order to control the stressing operation. The monitoring was continued to observe creep in the stressed concrete and also to control the subsequent re-stressing after an interval of a year.

The modulus of elasticity of the concrete, once the crack had been closed, was measured (via the strain gauges) as 10 GPa which agreed with the lower limit to values of the modulus measured on cores taken from the cantilever. During subsequent checking and adjusting of the post-stress, the loss of post-stress was used to estimate

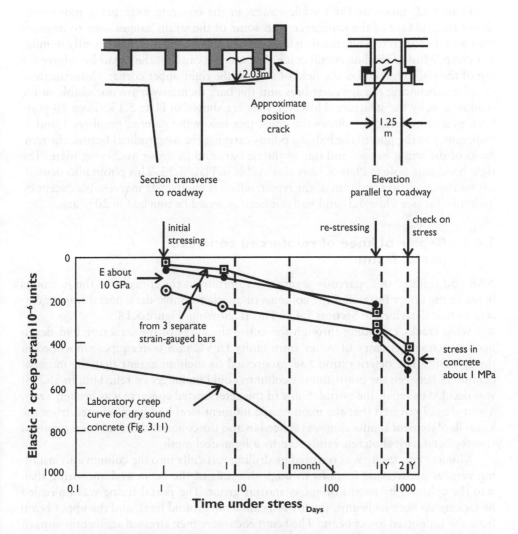

Figure 5.17 Details of dimensions (above) and measurements of field creep curve (below) for cantilever projection damaged by AAR.

the creep that had occurred in the concrete. Figure 5.17 shows the resulting field creep curve which is compared with a laboratory creep curve measured on a dry core of sound concrete drilled from the cantilever (See Figure 3.11). When the measurements were terminated after two years, the effective modulus of elasticity of the concrete had declined from 10 GPa to just over 2 GPa. As time has been shown on a log scale in Figure 5.17, it is not obvious that the field rate of creep strain actually declined from 160×10^{-6}/y in year one to 100×10^{-6}/y in year two. This should be compared with an overall creep rate measured on cores in the laboratory (Figure 3.11) of 1700×10^{-6}/y. (Note that Figure 5.17 is a corrected version of Figure 5 in the paper by Blight, 1990).

Plate 5.12, taken in 1983 while strains in the concrete were being monitored, shows the end face of the cantilever with some of the strain gauges used to monitor strains in the concrete while the tie bars were being stressed, and subsequently to monitor creep. Most of the horizontal crack that was the cause of the repair lies above the top of the photo but enters the field of view in the right upper corner. Unfortunately, the plate anchoring the stressing bars and the bars themselves, are not visible in the shadow cast by the structure. However, they are shown in Plate 5.13, taken 20 years later, in 2003. Plate 5.13 shows the crack just below the painted numbers 1 and 5, (top, centre in the photo) the halving points carrying the longitudinal beams, the remnants of the strain gauges and one of the tie bars and its lower anchoring plate. The right hand core hole in Plate 5.12 is also visible in Plate 5.13. This photo also demonstrates the excellent condition of the repair, which is completely inaccessible except by hydraulic lift (see Plate 5.5) and had not been accessed or touched in 20 years.

5.4.2 Repair of knee of reinforced concrete portal frame

AAR had resulted in apparently severe deterioration of the concrete at the junctions between the upper beam and the columns of another double-deck portal frame adjacent to that described in Section 5.4.1. This is shown in Figure 5.18.

Wide cracks extending through the entire thickness of the structure had developed as a result of entry of water from faulty UPVC rain water pipes embedded in the columns. The deterioration had progressed to such an extent that the moment continuity between the beam and the columns could no longer be relied on. In 1979 it was decided to repair the portal. Some of the deteriorated concrete was broken away, and it was discovered that the main tensile moment steel had been omitted from the knee. Replacement reinforcing was spliced in and the concrete that had been removed was replaced with shotcrete reinforced by a light steel mesh.

Moment continuity was restored by drilling vertically into the column and installing vertical prestressing tendons through the ends of the beam and anchoring then into the columns by means of epoxy mortar grout. The portal frame was unloaded by jacking up both its beams, the lower beam from ground level, and the upper beam from the supported lower beam. The beam ends were then stressed against the tops of the columns. Provision was made for re-stressing the tendons at suitable time intervals to make up for any loss of prestress that may have resulted from creep in the beam ends, the columns and the epoxy mortar-grouted anchorages.

The maximum design stress in the concrete at the junction of the column and the upper beam of the structure was 7.6 MPa on the inside of the column. A maximum stress of 2.1 MPa resulted at the outside of the column from the remedial prestrssing. The maximum design stress was, however, virtually unaffected by the prestress. The maximum design stress was still well within the strength of the deteriorated concrete. The tendons were re-stressed twice, after one and two years, when the loss of prestress was found to be less than 10% in each case.

Twenty four years later, as shown by Plate 5.14, photographed from a hydraulic lift in 2003, the repair looks almost as good as it appeared at completion. The only visible deterioration is some rust-streaking coming from the anchorages at the top of the knees and from the 250 mm square anti-bursting restraints just above the level of the top of the anchorages. (See Figure 5.18).

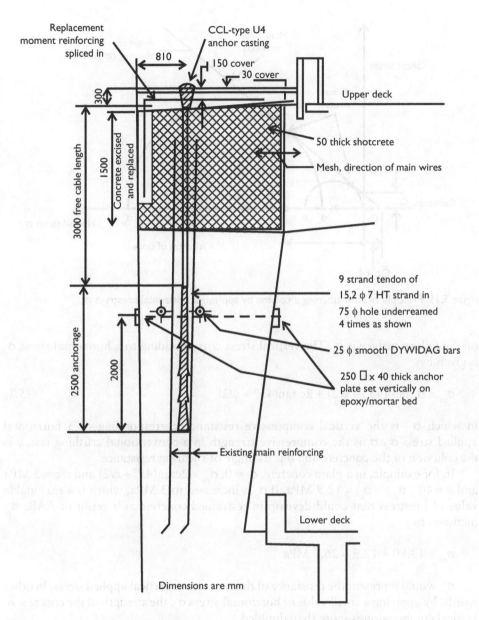

Figure 5.18 Knee of double-deck portal structure seriously damaged by the AAR, showing remedial measures.

5.4.3 Principle of increasing resistance to vertical stress by increasing horizontal stress

The principle is illustrated by Figure 5.19 which shows the Mohr failure envelope for concrete (see Figure 3.7), drawn in shear (τ) and normal stress (σ) space. The envelope is described by $\tau_f = c + \sigma_f \tan \phi$ where τ_f is the shear strength when the normal stress

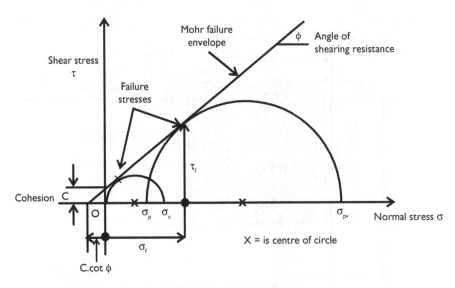

Figure 5.19 Principle of strengthening a column by applying a horizontal prestress σ_p.

on the failure surface is σ_f. The vertical stress corresponding to a horizontal stress σ_p is given by:

$$\sigma_{pv} = \sigma_p \tan^2(45° + \phi/2) + 2c \tan(45° + \phi/2) \tag{5.1}$$

in which σ_{pv} is the vertical compressive resistance corresponding to a horizontal applied stress σ_p. σ_c is the compressive strength in a conventional crushing test, c is the cohesion of the concrete and ϕ is its angle of shearing resistance.

If, for example, in a plain concrete, $\sigma_p = 0$, $\sigma_{pv} = 2c\tan(45° + \phi/2)$ and if c = 3 MPa and $\phi = 40°$, $\sigma_{pv}(= \sigma_c) = 12.9$ MPa. If σ_p is increased to 3 MPa, which is a reasonable value of prestress that could develop in restrained concrete as a result of AAR, σ_{pc} increases to

$$\sigma_{pv} = 13.80 + 12.9 = 26.7 \text{ MPa}$$

σ_{pv} would represent the resistance of the concrete to vertical applied stress. In other words, by applying a small value of horizontal stress σ_p, the strength of the concrete in vertical compression is more than doubled.

If an additional horizontal prestress of 3 MPa is applied, i.e. σ_p becomes 6 MPa, σ_{pv} becomes 40.5 MPa. Hence a very modest additional horizontal prestress applied to a confined concrete, swollen by AAR, can considerably increase its resistance to vertical stress, in this case, from 26.7 to 40.5 MPa, a factor of 1.51.

This explains very well, why prestressed concrete structures and structural components, see, e.g. Sections 4.11.1, 4.12.1 and 4.12.3, have performed so well over many years of service, despite being subject to continuing AAR. Even the conventionally reinforced flat slabs described in Section 4.12.2 were confined by their top and bottom reinforcing and by the edge beams, and therefore continued to have a satisfactory strength in shear. The concrete roadway described in Section 4.10.4 was also confined transversely by friction between the underside of the concrete and the surface of the road base, and longitudinally by the continuity of the carriageway.

The principle is a very old one, dating back to Rankine's work in the 1860s (Rankine, 1862) and has been widely used in many applications. An example is the use of horizontal reinforcing in the form of steel mesh, hoops or wire spiral wrapping in underground mine supports to increase their resistance to vertical loading (Hahn, et al., 1980, Blight, 2010). It also underlies the repeated observation, made as a result of structural load testing, that in situ strengths of AAR-affected concrete, confined by reinforcing, can be considerably larger than those of cores taken from the structure. The shear strength of the concrete is also increased. According to the Mohr failure envelope,

$$\tau_f = c + \sigma_f \tan \phi \qquad (5.2)$$

If $c = 3$ MPa and $\phi = 40°$ then if $\sigma_f = 0$, $\tau_f = 3$ MPa, but if $\sigma_f = 3$ MPa, $\tau_f = 5.5$ MPa, and if it increases to 5 MPa, $\tau_f = 7.2$ MPa.

The important feature of the Mohr theory is that it applies both to intact materials and materials consisting of discrete particles either interlocking, like fractured concrete, or non-interlocking like sand grains. In particular, concrete, disrupted by AAR and consisting of a mass of interlocking blocks would comply with the Mohr failure envelope under compressive stresses.

5.4.4   Strengthening column by means of stressed precast concrete encasement

Torii, et al. (2000) developed a means of strengthening an AAR-deteriorated column by wrapping the column with either a continuous stressed steel spiral or multi-start spirals, or stressed individual hoops. The method was used on the piers of the Toyokawa bridge, built in 1979, which was showing severe AAR damage. Each pier consisted of a pair of square piers measuring 2500×2500 mm that were joined by a vertical wall 800 mm thick. To repair the damage, the square columns of each pier were surrounded by segments of a 3900 mm diameter precast concrete cylinder as shown in Figure 5.20a. A spiral prestressing cable was then threaded into ducts cast into the precast cylindrical housing, and the spaces between the 150 mm thick cylinder and the square columns were filled with 40 MPa concrete. When the infill concrete had reached a strength of 30 MPa, the spiral was stressed to 30% of its yield strength. The arrangement of the original columns and the confining cylinders is shown in Figure 5.20b and also shows the square piers encased by the cylinders, as well as (b) the construction of the encasement.

The behaviour of the repair was monitored in detail for at least two complete annual cycles (730 days), as shown in Figure 5.20c. The traces in the two diagrams show that both crack width and strain in the repaired structure were responding to annual temperature cycles, but otherwise were changing by very little. It is possible, however, that irreversible strains could accumulate over several years and need further corrective action.

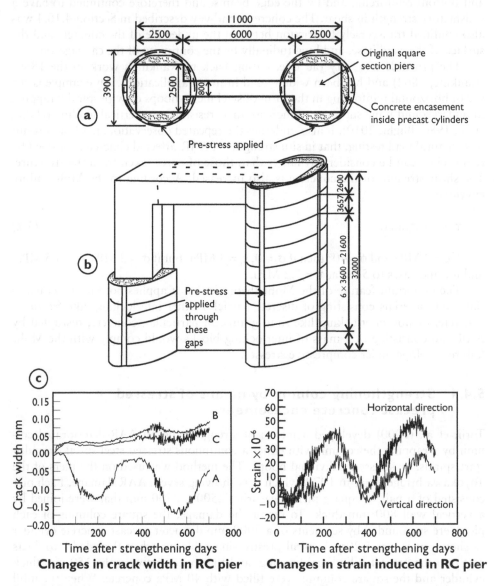

Figure 5.20 (a) Horizontal section of pier for Toyokawa bridge. (b) Elevation showing remedial encasement of square piers with prestressed cylinders (All dimensions are mm.). (c) Monitoring observations for 2 years following repair.

In a later paper, Torii, et al. (2004) reported that the repair has been very successful (6 years after its implementation) and that the stressed encasement method had been successfully applied to two more bridges at Omatagawa and Azumi. At Omatagawa, a reinforced road-under-road box culvert was strengthened by stressing a second supporting box culvert against it. At Azumi a free-standing square pier was strengthened by encasing it with an outer square precast encasement with a concrete infill, as at Toyokawa.

Wigum and Thorenfeldt (2004) have experimented with confining AAR-affected concrete by wrapping members with carbon fibre reinforced polymer (CFRP) sheets. Initial results appeared promising. However, the short-coming of this process appears to be that an active prestress (σ_p in equation 5.1) cannot be applied by means of an initially unstressed wrapping. The only way to induce a passive value of σ_p is to allow the AAR lateral swelling to progress and generate a value of σ_p as the wrapping of extremely high elastic modulus carbon fibres restrains the swelling. It would be more effective to apply the wrapping under controlled high tension.

5.5 STRENGTHENING BY GLUED-ON STEEL PLATES

Glued-on steel reinforcing straps and plates have been used for many years to augment the strength of highway bridges and other structures. For example, Plate 5.15 shows stainless steel stirrups glued to the underside of a reinforced concrete box girder bridge, damaged by AAR, to provide additional reinforcement against shear and torsion. Because structural glues creep under sustained stress, this additional reinforcing is usually used to provide only for dynamic or short-term live loading.

5.5.1 Experiments on external plating to strengthen concrete structures

Seto, et al. (2004) have reported on loading tests applied to very large scale specimens in a laboratory. The specimens represented double cantilever pier beams, and were 8.8 m long, 1.85 m high and 1.5 m thick. The dimensions (in mm) are shown in Figure 5.21. The problem was that the shear reinforcing in the prototypes which was in the form of stirrups, had fractured, but to an unknown extent. (See Sections 3.9 and 5.9 for information on causes of fracturing.) The tests were undertaken to see whether the beams could be strengthened by encasing the vertical sides and ends and the sloping soffits in a box made up of steel plate, bonded and prestressed to the existing damaged reinforced concrete. Three specimens were prepared:

1   An unplated reinforced specimen reinforced as for the prototypes and using 30 MPa concrete (E = 29 GPa). According to the paper, loading was to be applied at the one third points. (Figure 5.21 shows that the loading was actually over the central one third span (1750/5166)).
2   A fully plated specimen using 9 mm thick mild steel (f_y = 293 MPa), unreinforced weak concrete (13.4 MPa, E = 10 GPa), and a shear span greater than the flexural span.
3   A specimen identical to 2, except that the unreinforced concrete was even weaker (11.4 MPa, E = 9.6 GPa).

Specimens 1 and 3 had identical and unequal shear and flexural spans, Specimen 2 had a flexural span longer than the shear span, but the paper does not give actual dimensions. However, the central 1.75 m length of all three specimens was less than the cantilevers on either side, (again no width dimensions are given). The point loads seem to have been applied at the transition of the 1.5 m beam width to the wider central section.

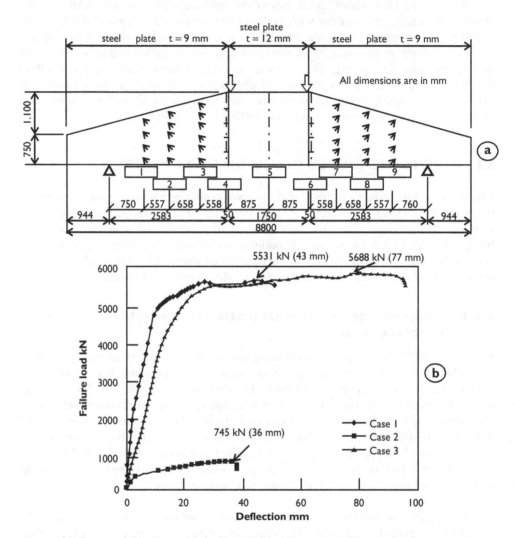

Figure 5.21 Tests on full-scale model double cantilever beams to test principle of strengthening by steel plate encasement: (a) Dimensions of beam (1500 mm thick) and positions (1 to 9) of lines of strain gauges. (b) Load-deflection curves in tests to destruction. The deflection at maximum load is given in mm for each test.

Figure 5.21 shows (a) an elevation of the test specimens showing dimensions and the location of the strain gauges rosettes (shown as arrowheads); and (b) the load-deflection curves for the tests to destruction on the three specimens. Figure 5.22 shows the distribution of strain measured at 4 sections along the beams (on the 1.5 m wide section next to the enlargement, and 558, 1216 and 1773 mm towards the support point. Strain measurements were seemingly only made on specimens 2 and 3.

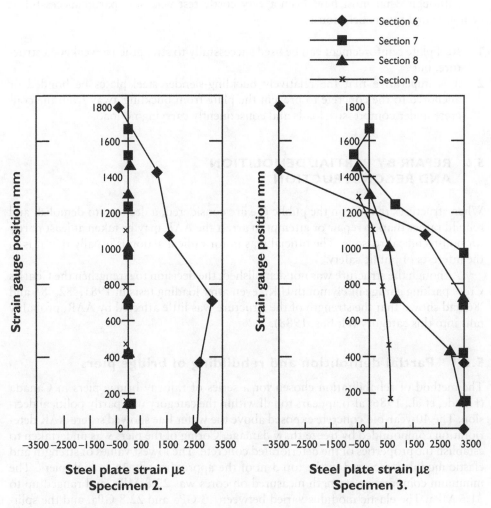

Figure 5.22 Relationship between strain gauge position and steel plate strain.

Specimens 1 and 3 gave almost identical load-deflection curves with ultimate loads of 5531 and 5688 kN respectively. Specimen 2 failed disastrously with an ultimate load of only 745 kN. Looking at the strain profiles for specimen 2 in Figure 5.22, all strains in the steel encasement were zero, except for the line of strains under the load point. This can only mean that the steel box was not bonded to the concrete nor effectively prestressed against it, and the plain concrete carried all the load. On the other hand, for specimen 3, the strains look more like what would be expected for plates bonded to the concrete, with a very high neutral axis in the steel. It appears that the steel probably yielded towards the lower end of the box ($f_y/E = 293/200 \times 10^3 = 1465$ microstrain, and strains well in excess of this were reached both in tension and compression).

Although what must have been a very costly test was only partly successful, it demonstrated two conclusions:

1   steel plate reinforcement can be used successfully to strengthen a weakened structure, but
2   it is imperative that the relatively buckling-slender steel plates be bonded or anchored to the concrete to prevent the plate from buckling away from the concrete under compressive loads and consequently carrying no load.

5.6   REPAIR BY PARTIAL DEMOLITION AND RECONSTRUCTION

When structures that are in the public eye are considered, a decision to demolish and rebuild rather than to repair or attempt to arrest the AAR may be taken at least partly for non-technical reasons. The official reason for radical action is usually that it is in the interests of public safety.

Although the structure was not demolished, the decision to strengthen the Charles Cross parking garage in Plymouth U.K., even after loading tests in 1981, '82, '85 and '86 had shown that the strength of the structure was little affected by AAR, probably falls into this category (Hobbs, 1988).

5.6.1   Partial demolition and rebuilding of bridge piers

The method of rehabilitation chosen for a series of railway bridge piers in Canada (Houde, et al., 1986) also appears to fall within the category of a partly political decision. The 40-year-old concrete exposed above the water line showed severe AAR deterioration compounded by freeze-thaw damage. Coring of the piers was undertaken to establish the properties of the deteriorated concrete. The lowest values of strength and elastic modulus occurred in the top 3 m of the approximately 16 m high piers. The minimum compressive strength measured on cores was 22.8 MPa, and ranged up to 31.6 MPa. The elastic modulus varied between 13 GPa and 22.8 GPa, and the splitting tensile strength averaged 2.5 MPa. The piers were massive, and these mechanical properties were probably adequate to support the applied loads. Also, the concrete was 40 years old and the rate of further deterioration must at that stage have been minimal. Nevertheless, a decision was made to demolish the piers above the water level and to reconstruct them. Improving the pier's appearance with a cosmetic treatment may have been as effective and much more economical a course of action.

5.6.2   Refurbishing a bridge underpass

Ryall, et al., (2000), describe a very interesting and intricate operation whereby a 38 year old road underpass to a two-span railway bridge and a second underpass beyond the railway in Sudbury, Canada, were refurbished and renewed by partial demolition and reconstruction. In this case the structures had been damaged by AAR, and the damage was compounded by freeze-thaw effects and corrosion of the reinforcing. Rehabilitation included removing and replacing the outer 300 mm of concrete and all of the reinforcing in the retaining walls. 200 mm of concrete were removed from the bridge abutments and replaced, and a central pier was treated by removing 75 to 100 mm of concrete (to 25 mm beyond the reinforcing steel). The objective of the rehabilitation was to gain at least an additional 20 year life from the structures. The description of the refurbishment was written less than 2 years after completion, so it is too early to judge the success of the operation. The refurbished structure certainly looks more handsome and therefore, in the mind of the public, safer than the cracked and squalid underpass it succeeded. Its working life has also probably been extended by 30 to 40 years.

5.6.3 Partial demolition and rebuilding of highway structure

The third example in this section concerns the partial demolition and reconstruction of the double decker portal frame previously test loaded in 1982 and 1988 (Section 4.10.2).

Severe surface deterioration of the western upper beam and column of the portal, specifically illustrated in Plates 4.6, 5.3 and 5.4, had been discovered in 1978. The portal had been monitored continuously from that time, and full scale test loadings were carried out in 1982 and 1988 (Section 4.10.2). These showed that the portal was able to carry its design load with a minimum of completely recoverable deflection and therefore behaved almost completely elastically under load. The consistency of behaviour of the portal in the two full-scale test loadings six years apart showed that the rate of deterioration of the structure was imperceptible over this period. Nevertheless, a decision based on non-technical considerations was made in 1991 to repair the portal by demolishing the upper beam and western knee (the cross-hatched portion in Figure 5.23) and rebuilding it.

As shown in Figure 5.23, the upper and lower beams were propped from a temporary foundation pad at ground level to support them, and the western portion of the beam, including the knee, was then demolished. Severe difficulty was experienced in breaking away the deteriorated concrete. Although the concrete had suffered AAR attack, its mean strength just before demolition was almost exactly the original design strength of 31 MPa (Alexander, et al., 1992). The reinforcing cage was found to be in perfect condition, without a sign of corrosion and, after breaking away the concrete, the original reinforcing cage was re-used, just as it was. Plate 5.16 shows a stage in the demolition with the perfectly preserved main reinforcing exposed after 27 years.

The replacement concrete used a dolomite aggregate known to be free of potential for AAR, and as a further precaution the equivalent Na_2O content of the cement was limited to 0.5%, with a total cement-alkali content of 1.4 kg/m³. Test prisms gave a mean compressive strength of 55 MPa at 28 days and an elastic modulus of 46 GPa, also at 28 days.

When the repaired beam was de-propped at 28 days, it sagged by 2.6 mm at the junction between old and new concrete and a few cracks formed at the junction (Blight, et al., 1993). Strains and cracking on the surface of the beam were monitored for a year after the de-propping (see Figures 5.24 and 5.25). The cracks remained at hairline width and maximum concrete surface strains were 600 microstrain in tension and 200 microstrain in compression. Five and a half years later (in 1997) strains

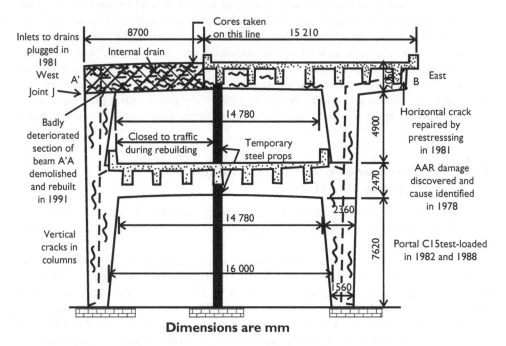

Figure 5.23 Main dimension of double deck portal frame, showing history of repair prior to upper beam being partly replaced in 1991 (elevation looking north).

on the beam were checked (Blight and Lampacher, 1998). It was found that the new concrete had shrunk by about 500 microstrain relative to its condition in 1991, but that the old concrete had expanded by almost 1000 microstrain in the same period. The strains measured in 1997 are shown as contours in Figure 5.25. Cracking at the junction of the new and old concrete, however, remained very minor. Sketches of the cracks, made in 1992 are shown in Figure 5.24 and in 1997, in Figure 5.25. These can be compared with Plate 5.17, a close-up photograph taken in 2003, and show very little, if any, change in 11 years.

Plate 5.17 shows the junction between the old and new concrete in 2003, with the remains of the "water-proofing" coating applied to the original beam in 1989 (Section 5.2.3) to the left and the Demec targets applied before de-propping the reconstructed beam in 1991. The gauge length of the Demec targets is 200 mm, applied to form 0°, 45°, 90° rosettes. A small area of the original AAR-cracked concrete is visible above the 100 mm length of steel tape stuck to the concrete half-way up the picture. The cracks in the new concrete occurred when the beam was de-propped. Plate 5.18 is a view of the complete rebuilt beam taken in 2003, 12 years after its reconstruction.

In October 2003 the opportunity arose to inspect the beam from close quarters from a hydraulic lift. The Demec targets originally used to measure surface strains were still intact, but, unfortunately, the zero measurements had been lost and hence strains that had occurred since 1997 could not be measured. The cracking was found to have changed very little over the six years since 1997 and the opportunity was

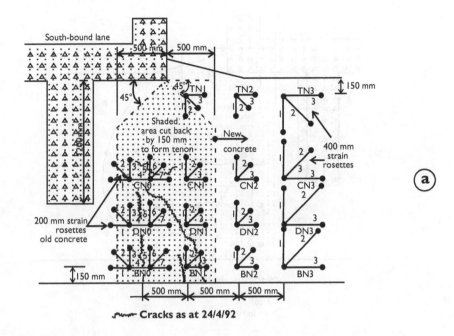

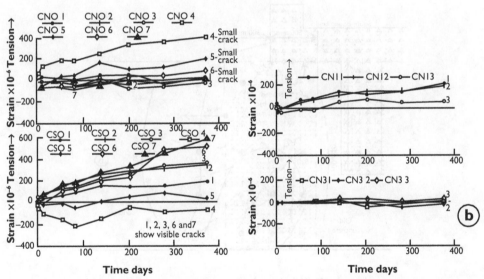

Figure 5.24 Reconstruction of beam of portal, damaged by AAR. (a) Arrangement of Demec strain gauge rosettes in the vicinity of the tenon joint between old and new concrete. (b) Strains measured on surface of repaired beam over one year following repair.

taken to establish a new set of strain zero readings. Strains were re-measured in October 2004 and 2005, but only changes that could be ascribed to the ambient temperature differences between the three sets of measurements were detected. Apart from the slight increase in surface cracking, the repair was in perfect condition.

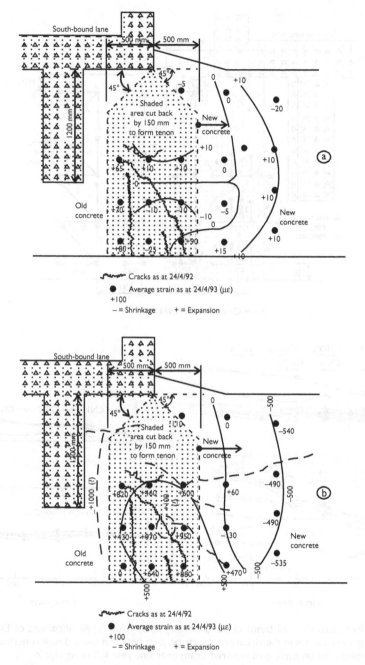

Figure 5.25 Average strain contours on reconstructed beam. (a) Average strains and cracking recorded a year after the repair. (b) Average strains and cracking recorded 5½ years after the repair.

5.7   REPAIR AND REHABILITATION OF CONCRETE HIGHWAY PAVEMENT

The 200 mm thick unreinforced concrete highway pavement damaged by AAR, that was described in Section 4.10.4 had been built in 1969. Hairline cracking in the vicinity of the sawn joints was noticed in 1975 and this was diagnosed as the result of AAR. By 1979 loose blocks had developed adjacent to the joints (see Plate 4.15 and Figure 4.20). A study was initiated in 1980 to consider possible remedial actions.

Observations had shown that water entering the sub-base and sub-grade (Figure 4.19) through the cracks was reducing the slab support, while the AAR expansion, restrained by friction between the slab and the base resulted in high horizontal compressive stresses developing in the concrete, closing undeteriorated joints and restricting thermal movements. Horizontal compressive stresses were calculated, from stress-relieving tests, to be about 4 MPa. (Compare this with the range of swelling pressures in structures of 2 to 4 MPa arrived at in Sections 3.8.2 and the 0.4 MPa estimated in Section 5.8.1).

St. John (1986) reported damage to a concrete pavement at an airforce base in New Zealand. The damage occurred by upward buckling at the joints of abutting slabs. He ascribed this to horizontal swelling pressure developed by AAR. However, such "blow-ups" occur in concrete (and even brick and asphalt) pavements where there is no evidence or possibility of AAR, purely as a result of restrained thermal expansion. In the case reported, AAR may have been an aggravating factor, but is unlikely to have been the primary cause of the damage.

Three experimental repair methods for the joints resulted in the conclusion that the best method was to cut out the cracked concrete, leaving jagged edges to enhance load transfer and bond, with a high quality replacement concrete used to reconstruct the joints. In situ measurements showed that this resulted in a slight reduction of relative movement across the joint from mean ($\bar{x}$) and standard deviation (s_d) values

from $\bar{x} = 0.102$ mm $s_d = \pm 0.065$ mm
to $\bar{x} = 0.090$ mm $s_d = \pm 0.010$ mm

The main improvement was thus in the consistent quality of the replaced joints.

The pavement, with reconstructed joints, was then overlaid with an asphalt-rubber membrane as a waterproofing measure and a 45 mm thick asphaltic concrete overlay with a bitumen-rubber binder on top. The expected life of this repair, which was laid in 1986, was 7 years, after which it was planned to replace the asphalt with a continuously reinforced concrete pavement. However, because of their noisiness and indifferent ride, concrete road pavements have gone out of favour in South Africa, and the road has not been reconstructed, but continues in service with an asphalt surface 24 years later (Strauss and Schnitter, 1986).

5.8   REPAIR OR MITIGATION OF EFFECTS OF AAR IN LARGE MASS CONCRETE STRUCTURES

In large structures constructed of mass concrete, for example, gravity dams, gravity arch dams, underground powerhouse linings, etc. small strains caused by AAR expansion can accumulate to cause relatively large movements and distortions at openings between large structural blocks and in the alignment of shaft bearings, etc. Because most large dams are situated in remote areas and are not frequently seen by the public, it is not generally known how frequently AAR problems occur in this type of structure. Brazil will be taken as an example of problems that may occur in mass concrete dams attacked by AAR. With its large rivers, Brazil probably has more large mass concrete hydraulic structures than any other single country in the world. As of 1999 (Andriolo, et al.) Brazil had 19 large dams that exhibit AAR problems (large dams are defined as dams over 30 m high) out or a total number of 830. (A small proportion of 2.3%, but nevertheless a large number.) (The 830 dams include many earth and rockfill dams, so the proportion of concrete dams with AAR damage is much larger than 19 in 830.) The dates of completion vary from 1926 to 1979 and the symptoms of AAR damage include cracks, opened construction joints, displacements between blocks, movement of dam crests and misalignment of turbines, usually caused by differential movement of bearing blocks. All of the dams that exhibit three or more of these symptoms were constructed in the period from 1971 to 1979, and the damage was first noted in 1996 for the dams completed in 1971 and 1985 for the one completed in 1979. Measured expansion rates have varied from 13×10^{-6} per year for a gravity dam completed in 1964, and which exhibits cracking, to 90×10^{-6} per year for another gravity dam completed in 1977 and exhibiting all five of the problems listed above.

Plate 5.19, showing the Khatse arch dam under construction, in Lesotho, illustrates the many opportunities for AAR damage to occur in these mega-size multi-blocked, multi-lift structures. The Khatse dam, completed in 1996, is 185 m high and has a crest length of 710 m. Its double curvature structure contains 2 320 000 m³ of concrete and was constructed in 32 interlocking segments each of which was built in many lifts of concrete. Natural variations in the huge quantities of aggregate, both coarse and fine, variations in the properties of the cement used, often coming from several suppliers, and variations in the resultant concrete mix, all provide possibilities for problems such as AAR or DEF (Delayed Ettringite Formation) to occur.

Some of the remedial measures and repair methods that have been used in this type of structure, in various parts of the world, will now be described.

5.8.1   Use of slot-cutting to relieve distress in hydroelectric power machinery

Silveira, et al., (1989) illustrate some of the problems related to hydro-electric turbines when their supporting concrete foundations become distorted or displaced by pressures generated by AAR. These are illustrated generically by Figure 5.26. The numbers on the section through a typical hydroelectric turbine refer to:

1   friction between turbine blades and discharge rings;
2   ovalling of discharge rings;
3   mis-alignment of the stay vanes;
4   ovalling of the generator stator;
5   ovalling of the guide bearings; and
6   mis-alignment of the turbine/generator axis.

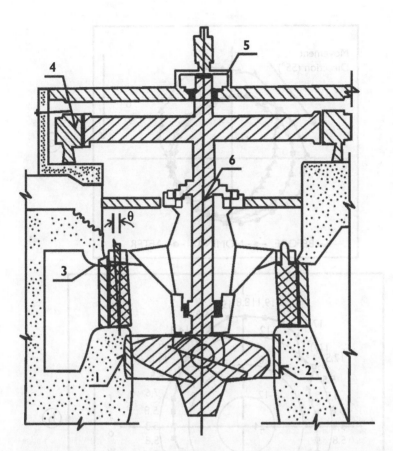

Figure 5.26 Section through typical hydro-electric turbine showing points at which distortions caused by AAR in the concrete housing can result in distress: 1: turbine blade; 2: discharge ring; 3: stay vane 4: generator stator, 5: guide bearing, 6: turbine/generator shaft.

The usual way of overcoming the effects of these distortions is by cutting slots through the concrete in strategic places and directions to free the constraints and forces developed in the concrete, which then act on the machinery.

Cavalcanti, et al. (2000) give a detailed exposition of the operations necessary to restore one of the Kaplan turbines at the Apolonia Sates hydro-power station in northern Brazil to good working order. Some of the distortions that resulted in the need for repair are illustrated in Figure 5.27. Referring to Figure 5.27a, the tops of the runner blades are designed to clear the discharge ring by 6 mm all round. The required 6 mm clearance is shown by the circle marked "nominal" in the figure which, to scale, is 6 mm all round. Before the intervention, this had closed to only 3.3 mm over part of the ring's circumference and opened to 8.7 mm elsewhere. Figure 5.27b shows the total diametral gap, i.e. the sum of diametrically opposite clearances. This should be 12 mm all round, but varied from 5.3 to 19.2 mm. Obviously, the work, deep underground, required to restore these clearances to close to the specified values, was extremely arduous and technically difficult. It comprised cutting away the concrete

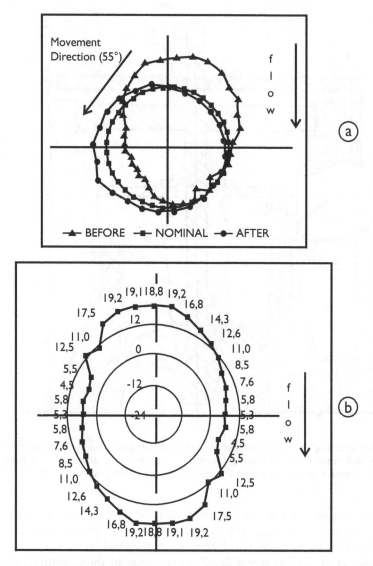

Figure 5.27 Some of the deformations caused by AAR that necessitated remedial measures to the casings of the Kaplan turbines at Apolonia Sales hydro-power station in Brazil. (a) Relative position between the turbine runner and discharge ring. (b) Runner – discharge ring diametral gaps (mm).

where it had caused intrusion on clearances and realigning shafts and bearings so that the turbines could run without hindrance for several more years.

The complexity of these structures and the difficulty of deciding where to cut the stress-relieving slots is illustrated by Figure 5.28 which compares the movement contours predicted by a three-dimensional finite element analysis with spot

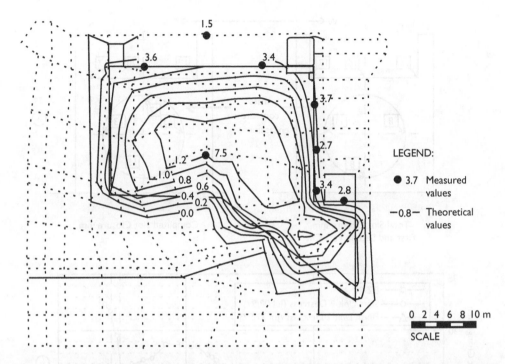

Figure 5.28 Comparison of predicted movement contours in a concrete turbine housing as a result of stress-relieving slot cutting, with spot measurements of movements that actually occurred.

measurements of the movements that actually resulted (Silveira, et al., 1989). In particular, the values of 3.6, 3.4 and 3.7 mm on the predicted 0.0 mm movement contour and 7.5 mm on the 1.2 mm contour illustrate the difficulty of making an accurate prediction on which to base remedial slot cutting.

In a related, but less geometrically complex problem, slot cutting was described by Curtis (2000) as a means to relieve distress in the concrete encasement of a penstock housing. Figure 5.29a compares (left) measured movements following slot-cutting with (right) computed movements. In this (less geometrically complex) case, agreement between computation and measurement was quite good. Figure 5.29b shows movements of the generator floor related to the slot cutting. Slot cutting is recognized to be a load relieving procedure that will need to be repeated periodically as swelling caused by AAR proceeds. In this case the movement induced by the slot cutting (between the vertical dashed lines in Figure 5.29b) was dramatic, but the general movement continued after slot cutting was completed. The total compressive force on the penstock caused by AAR expansion was estimated at 50 MN. The diagrams in the paper are not dimensioned, but working from the size of the penstocks (8.84 m internal diameter) and scaling from Figure 5.29a, the cross-section of the concrete is about 120 m² and hence the compressive stress was about 0.4 MPa. This falls well within the range of AAR swelling pressures of up to 4 MPa estimated in Sections 3.8.2 and 5.7 in structures

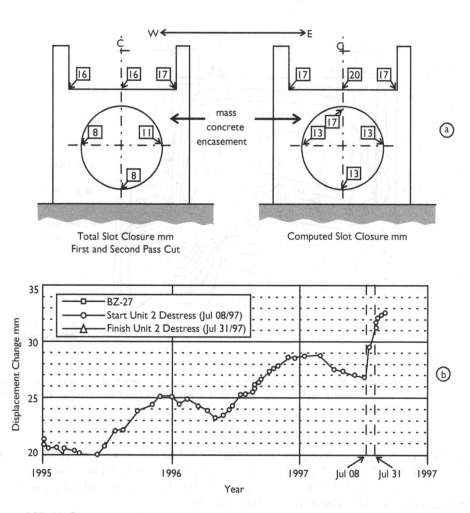

Figure 5.29 (a) Comparison of predicted and measured movements caused by stress-relieving slot cutting in a mass concrete penstock encasement (Curtis, 2000). (b) Movements caused by AAR in the mass concrete penstock encasement referred to in a), showing the effect of slot cutting.

exposed to the weather. (In this application, the concrete, being underground, is not exposed to desiccation and stays at a reasonably constant temperature).

5.8.2 Effects of AAR on movements of arch dams

The following summary has been made from a review of the effects of AAR on arch dams with a range of heights, lengths and ages and located in many parts of the world, by Curtis (2000).

- Observed amounts of vertical expansion strain obviously depend on many factors and vary from a low of 6 to 28 × 10^{-6} per year in the highest dam reviewed, the

170 m high, 300 m crest length Cahora Bassa dam in Mozambique in which the compressive stress and hence restraint imposed by the great height of concrete must play an important role, to highs of 100 to 120×10^{-6} per year in the 40 m high by 131 m crest length Gene Wash dam in California. Intrinsic differences in expansiveness, aside, the concrete in a higher dam would be more restrained by its self-weight and therefore would tend to less average vertical expansive strain, although the actual expansive movement at the crest of the dam might not be less. (For example: $28 \times 10^{-6} \times 170$ m = 4.8 mm/y for Cahora Bassa, and $120 \times 10^{-6} \times 40$ m = 4.8 mm/y for Gene Wash.

- Because the height of concrete varies from midstream (maximum height) to the abutments, less expansive movement may occur near the abutments than at midstream, and this is thought to have caused diagonal cracking near the abutments of some arch dams (e.g. Gene Wash dam and the 57 m high, 77 m crest length Copper Basin dam in California). The inset on Figure 5.30 shows measurements of the increase in height with time of two dams, one of which, the Copper Basin dam, recorded an average expansive strain of 1600×10^{-6}. As the height is 57 m, the increase in height is 0.09 m, or 90 mm.

- Because horizontal swelling is restrained by the horizontal arch thrusts between the abutments, arch dams subject to AAR will be subject to forced horizontal upstream movement. For several arch dams (e.g. the 78 m high, 317 m crest length Kouga dam in South Africa and the 76 m high, 115 m crest length Santa Luiza dam in Portugal), the distribution of horizontal movement is shaped like an "M", with more horizontal movement occurring at the quarter span points than at mid-span.

Figure 5.30 (main diagram) shows measurements of the heights of the multiple arch-buttress Churchill dam in Port Elizabeth, South Africa. (Oberholster, 1989 NBRI, 1963). (The "saw teeth" in the height measurements arise because of annual seasonal temperature variations). The Churchill dam was built between 1940 and 1943 and consists of 10 inclined arches spanning 17 m between mass concrete buttresses: The overflow crest height is 36.5 m. The circular arches are inclined downstream at 60° to the horizontal, and taper in thickness from 3.4 m at the base to 1.8 m at the crest. The arches carry hoop reinforcing in both faces, but are only nominally reinforced parallel to their axes.

Horizontal cracking was first observed on the downstream face of some of the arches in 1957, and level measurements have shown increases in height of the arches of up to 26 mm in the 20 years between 1960 and 1980 (Arch NOA3 in Figure 5.30), equivalent to an average strain rate of 35×10^{-6} per year. Since 1984 the rate of expansion has slowed almost to zero (Figure 5.30), but the overall expansive strain is at least 1500×10^{-6}. In 1960, 17 years after completion, and 27 years later in 1987, cores were taken from the dam and compressive strengths and elastic moduli were measured. The results of these tests were:

Core taken in	Compressive strength σ_c	Elastic modulus E
1960	30–44 MPa	16–31 GPa
1987	28–50 MPa	20–33 GPa

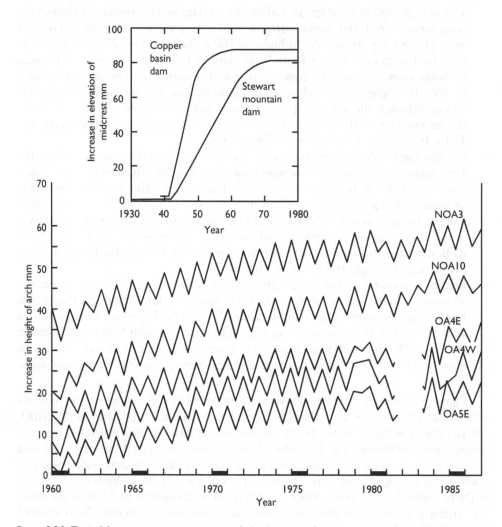

Figure 5.30 Typical long-term measurements of the heights of arches of the Churchill Dam, Port Elizabeth, South Africa. (Inset are curves recording the height expansion of two relatively low arch dams. The 57 m high Copper Basin dam is particularly interesting as it expanded in height by a strain of 1300×10^{-6} in less than a decade.).

These measurements were very re-assuring, although the usual reservations concerning the effects of selectivity in choosing the sites from which to core must be borne in mind. The Churchill dam continues in operation and according to the owners, Port Elizabeth Municipality, is in a satisfactory state.

5.8.3 Slot-cutting for relief of swelling stress

It might be expected that if a slot were to be cut to relieve swelling stress in a large mass of concrete affected by AAR, then unless the slot is very wide, it would gradually close up, because cutting a slot will not stop continued swelling of the concrete but may initially accelerate it because of the removal of restraint. It should also be noted that in the clearance cutting exercises described in Section 5.8.1, there was no expectation that the remedial measures would be permanent, but only that the turbines could run freely again until continued swelling of the concrete necessitated a repeat of the exercise.

A review of experience with slot-cutting in concrete dams affected by AAR in Canada (Gocevski and Pietruszcak, 2000) lead to almost the same conclusion. An example of their data is shown in Figure 5.31, and is very instructive: Figure 5.31a

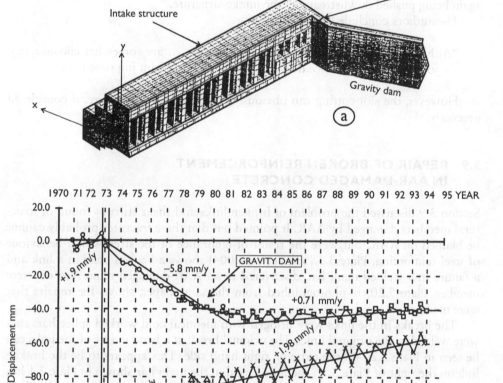

Figure 5.31 Intake structure and gravity dam in Canada. Intake structure expanding and pushing gravity dam in downstream direction. (a) Intake structure and dam. (b) Movements prior to and after cutting stress-relieving slot at juncture of structures in 1972–73.

shows a concrete gravity dam that abuts a massive concrete water intake structure built at right angles to it. Expansion of the intake structure was causing the upper part of the dam to be pushed downstream. The movements of both structures are recorded in Figure 5.31b and show that (as in Figure 5.30) a seasonal, mainly temperature movement is superimposed on the overall movements. Once a slot was cut in 1972/75 to free the two structures from each other, the top of the dam rebounded steadily in an upstream direction by 42 mm until, in 1981, the slot was closed by the accelerated swelling of about 35 mm by the intake structure, and 7 mm by the dam. Once the slot had swelled shut, the accelerated movement of the intake structure slowed, and the direction of movement of the dam reversed. The two structures were, by 1994, moving towards the original equal rates of movement of about +1.9 mm/y with the dam again being pushed downstream by the intake structure.

The authors concluded that:

"Although the immediate results of slot-cutting .... are somewhat encouraging, the long-term performance does not show any significant improvement".

However, the slot-cutting can obviously be repeated in the future, if considered necessary.

5.9 REPAIR OF BROKEN REINFORCEMENT IN AAR-DAMAGED CONCRETE

Section 3.9 discussed the problem of broken links and shear stirrups found in structural members damaged by AAR. It pointed out that these fractures probably cannot be blamed on AAR, but were the result of ignorance of the stress-strain behaviour of steel reinforcing. Plate 3.1 (Seto, et al., 2004) shows a broken bend in a link and a failure in a welded main bar, discovered in an AAR-damaged reinforced concrete member. Plate 5.20, re-photographed from the same paper, shows the repairs that were made:

The breaks in the links were bridged with identical cold worked splice bars that were welded to the original links. The original break at a bend in one of the links can be seen in Plate 5.20 on the extreme right hand side. This appears to be the broken link on the right of Plate 3.1. The failed weld in the main bar shown in Plate 3.1 has been repaired by cutting out a short piece of bar and welding in a splice length. There also appears to be a welded-in splice length in the next longitudinal main bar from the corner.

It is most unlikely that these repairs will have succeeded, as they merely repeated the faults that originally resulted in fracture of the reinforcing. The spliced-in bars, both 10 and 16 mm diameter are again of cold worked steel, the links have been bent to similar radii, and instead of one weld in the 16 mm bar, there are now two. The authors did not realize that two negatives do not yield a plus.

In their paper, Nomura, et al. (2004) concluded: "... the main cause of damage is not the poor quality of rebar itself but that expansion of the concrete of ASR gives an excessive stress on weaker bending parts." Seto, et al. (2004) do not disagree with this. Kuzume et al. (2004), however, put their finger on the real problem.

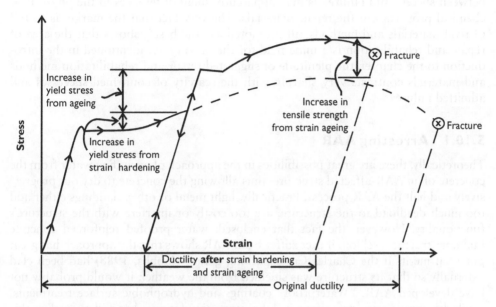

Figure 5.32 The concept of strain ageing in steel.

To quote: "When bars are bent, there is a chance to cause cracks in(side) the bent area if the bending radius is rather small." (Posssibly the authors meant in the bent area). There is only one possible thing wrong with this observation, and this is that cracks will not originate on the inside of the bend, but on the outside where strains are both tensile and largest (see Figure 3.12b). Kuzume, et al. also state that strain ageing reduces tensile strength. However, according to the literature, strain ageing, in a temperature range from 15 to 30°C occurs within a few days of straining the steel to beyond the yield strain. Strain ageing has the effect of slightly increasing the yield stress and the ultimate tensile strength, it does not decrease the strength. (South African Institute of Steel Construction, 2010.) Figure 5.32 illustrates the phenomenon of strain ageing in steel.

The question arises of how the repairs should have been made. There are mechanical clamps of several types on the market for the full-strength joining of reinforcing bars. These would have been a better choice than welding. Using mild steel splices clamped on the links, of a larger diameter, if necessary, and bent to a more suitable radius, would also have been a better choice than to weld on cold worked bars.

5.10 REVIEW AND CONCLUSIONS

The conclusions set out below should all be viewed as tentative. The body of published information on methods of repair and rehabilitation, for which success has been confirmed by several years of post-repair observation and/or measurement, is sparse. A small detail in application, such as the method of surface preparation or the weather at the time, or prior to making a repair, may make the difference between success and failure, as may apparently small differences in the physical or chemical properties of the repair materials. The very fact that the market is so full of rival materials and methods, all promoted as "the best", shows that the area of repair and rehabilitation is a mine-field for the unwary. As mentioned in the introduction to the chapter, the plenitude of suggested repair and rehabilitation methods and materials contrasts very sharply with the scarcity of confirmed successes and admitted failures.

5.10.1 Arresting AAR

Theoretically, there are great possibilities in the approach of excluding water from the concrete of an AAR-affected structure, thus allowing the concrete to dry out progressively and halt the AAR process. Practically, light metal or other claddings either add too much deadload to the structure, are too costly or interfere with the structure's functionality. However, the fact that enclosed, water proofed reinforced concrete building structures seldom if ever suffer from AAR shows that the approach has great potential merit. If the Charles Cross parking garage (Hobbs, 1988) had been clad externally so that its structure was shielded from the weather, it would probably not have developed AAR. "Waterproof" coatings and hydrophobic surface treatments, similarly, appear very promising in laboratory simulations, but fail to realize this promise in practice. There seems nearly always to be an unrealized or hidden defect that lets water into the concrete, even if the protective surface treatment performs as claimed by its supplier (see Section 5.2.3).

The use of lithium compounds to arrest or suppress AAR does not yet appear to be sufficiently well tried to be recommended.

5.10.2 Repair by resin injection

There appear to be very few examples of this technique that have been reported in the literature. Initial trials by the authors did not prove successful and the method appeared on theoretical grounds to be unpromising. Success was claimed for the Hanshin expressway structures, but this was based on UPV measurements which were unlikely to have been a reliable basis for the claim. However, the success of the repair of a sports stadium (Section 5.3.2) in South Africa has been confirmed by a series of careful measurements from 1992 to 2002 and is likely to be further confirmed when the next set of measurements (in 2012) is made.

This is a method that warrants further exploration and exploitation.

5.10.3 Repair by externally applied stressing

Two examples of repairs in South Africa, confirmed to be successful over a period of 20 years have been reported in Section 5.4 and a variant of the method, applying lateral stressing to strengthen against vertical stress, has been reported successful in Japan. The first use of the method in Japan (Section 5.4.4) has been confirmed as successful over a period of at least 6 years and has been applied in at least three cases.

5.10.4 Repair by external reinforcing

This method has been experimented with in Japan using external plating, but was not successful, mainly because of what appears to have been a poorly planned experiment. It would be well worthwhile trying the method again, this time with a carefully analysed and performed experiment. It should also be noted that other forms of this method, using glued-on strips rather than plates (see Plate 5.15) have proved very effective.

5.10.5 Partial demolition and reconstruction

This is the fall-back solution if all else fails, or if the situation is such that it is too risky to attempt other measures, or it may be more economical to undertake. It has the special merits of allowing the engineer to correct shortcomings in the original structure, to remedy present or past aesthetics or even to extend the structure if extension is needed. The major difficulty is that of matching a new concrete structure with an old one. The three examples discussed in Section 5.6 were all reportedly successful.

5.10.6 Repair and rehabilitation of concrete pavements

Two examples have been dealt with. There may be other and better examples in the literature. However, all in all, asphalt does seem to be a better material as a road and aircraft runway surfacing than is concrete, and perhaps the concrete industry should accept this and concentrate on applications of concrete to pavements such as loading areas (Section 4.10.6), hard-standings for aircraft, toll plazas, shipping container stacking yards, etc. where riding quality is not important, but resistance to creep under static load and to damage by spilled hydro-carbons is important.

5.10.7 Alleviation of AAR effects in mass concrete structures

This is a very special problem that affects extremely large and costly structures that cannot easily be put out of service for repair. The measures used, such as slot cutting, therefore alleviate problems caused by AAR swelling in the full knowledge that further alleviation will probably be needed in future.

5.10.8 Broken reinforcement

When this problem is examined in the light of the well-known behaviour of strain-hardened, low-ductility steel, it becomes clear that it is a problem of reinforcing detailing and not a problem peculiar to concrete subject to AAR.

5.10.9 Repair and ongoing maintenance

As stated at the end of Chapter 4, the repair of AAR-damaged structures should be regarded in the same light as preventive maintenance of an undamaged structure.

Using humans (or any animal) as an analogy, preventive health care may be more expensive if the subject suffers from a congenital problem, but the life-span may still be similar to that of a congenitally normal subject, and the quality of life may remain excellent throughout the life-span. What is needed is an appropriate life-time maintenance plan. The type of plan for a particular structure must obviously be related to the degree of AAR damage and the consequences of failure, either by structural collapse or loss of functionality. One suggested scheme is to classify the AAR damage of each assessed structure into various grades, as follows:

Low level:	structure with little damage and slow or intermittent AAR development. : or structure with extensive damage but with apparently no further AAR development, because of late stage reached by AAR process.
Medium level:	structure with more extensive damage but with slow or intermittent AAR development.
High level:	structure with little damage but with relatively rapid AAR development.
Ultra-high level:	structure with extensive damage and with rapid AAR development.

Once immediate repair measures have been taken, to the degree considered appropriate, structures could be inspected and assessed for further necessary repair on a schedule similar to the following:

Grade of damage:	Periodic examination to take place every:
Low level	5 to 10 years
Medium level	3 to 5 years
High level	Annually

There should be no ultra-high level of damage, as this should have been converted at least to High level after the initial corrective measures or repairs. Depending on what is found on the first few annual or triennial inspections and assessments, a structure could be down-graded from High level to Medium level to Low level, or vice-versa, it could be appropriately upgraded.

REFERENCES

Alexander, MG, Blight, GE & Lampacher, BJ 1992, 'Pre-demolition tests on structural concrete damaged by AAR', *Proc. 9th Int. Conf. on AAR in concrete.* London, United Kingdom, vol. 1, pp. 1–8.

Andriolo, FR 2000, 'AAR: dams affected in Brazil, Report on the current situation', *11th Int. Conf. on AAR in Concrete*, Quebec City, Canada, pp. 1243–1252.

Bakker, J 2004, 'Monitoring of ASR expansion and moisture in concrete', *12th Int. Conf. on AAR in Concrete.* Beijing, China, pp. 1202–1209.

Blight, GE 1966, 'Strength characteristics of desiccated clays', *Journal Soil Mech. and Found. Div. ASCE*, vol. 92, no. SM6.

Blight, GE 1990, 'Rehabilitation of reinforced concrete structures affected by alkali-silica reaction', *Structural Engineering Review*, vol. 2, pp. 113–120.

Blight, GE 1991, '(a) A study of four waterproofing systems for concrete', *Magazine of Concrete Research*, vol. 43, no. 156, pp. 197–203.

Blight, GE 1991, '(b) The moisture condition in an exposed structure damaged by alkali-silica reaction', *Magazine of Concrete Research*, vol. 43, no. 157, pp. 249–255.

Blight, GE 2008, 'An unexpected observation after drying AAR-affected concrete', *13th Int. Conf. on AAR in Concrete*, Trondheim, Norway, pp. 504–511.

Blight, GE 2010, 'Geotechnical Engineering for Mine Waste Storage Facilities', *CRC Press/Balkema*, Leiden, The Netherlands.

Blight, GE & Mitchell, MF 1980, 'The properties of epoxy resin mortars for use in nosings to bridge expansion joints', *Civil Engineer in South Africa*, 8, pp. 203–210.

Blight, GE, Alexander, MG & Lampacher, BJ 1993, 'Structural repair of reinforced concrete portal frame', *Magazine of Concrete Research*, vol. 45, no. 163, pp. 97–101.

Blight, GE & Lampacher, BJ 1998, 'Repair of reinforced concrete portal frame damaged by alkali-silica reaction – strains after 5½ years', *Magazine of Concrete Research*, vol. 50, no. 4, pp. 293–296.

Cavalcanti, AJCT, Campos, AT, Silveira, EMM & Wonderley, EG 2000, 'Rehabilitation of a generating unit affected by alkali-aggregate reaction', *11th Int. Conf. on AAR in Concrete*, Quebec City, Canada, pp. 1253–1262.

Curtis, DD 2000, 'A review and analysis of AAR-effects in arch dams', *11th Int. Conf. on AAR in Concrete*, Quebec City, Canada, pp. 1273–1282.

Delaby, J-B, Brouxel, M & Pascal, R 2004, *12th Int. Conf. on AAR in Concrete*, Beijing, China, pp. 1215–1218.

Gocevski, V & Pietruszczak, S 2000, 'Assessment of the effects of slot-cutting in concrete dams affected by alkali-aggregate reaction', *11th Int. Conf. on AAR in Concrete*, Quebec City, Canada, pp. 1303–1312.

Godart, B, Michel, M & Fasseu, P 1996, 'Treatment of structures by waterproof coating', *10th Int. Conf. on AAR in Concrete*, Melbourne, Australia, pp. 583–590.

Gudmundsson, G & Asgeirsson, H 1983, 'Parameters affecting alkali expansion in Icelandic concretes', *6th Conf. on AAR in Concrete*, Copenhagen, Denmark, pp. 217–222.

Hahn, JA, Dison, L & Blight, GE 1983, 'Supports of reinforced granular fill', *Int. Symp. on Mining with backfill*. Lulea, Sweden, pp. 300–306.

Hobbs, DW 1988, *Alkali-Silica reaction in concrete*, Telford, London, U.K.

Hooper, RL, Nixon, PJ & Thomas, MDA 2004, 'Considerations when specifying lithium admixtures to mitigate the risk of ASR', *12th Int. Conf. on AAR in Concrete*, Beijing, China, pp. 554–563.

Houde, J, Lacroix, P & Morneau, M 1986, 'Rehabilitation of railway bridge piers heavily damaged by alkali-aggregate reaction', *7th Int. Conf. on AAR in Concrete*, Ottawa, Canada, pp. 163–70.

Imai, H, Yamasaki, T, Maehara, H & Miyagawa, T 1986, 'The deterioration of alkali-silica reaction of Hanshin expressway concrete structures – investigation and repair', *7th Int. Conf. on AAR in Concrete*, Ottawa, Canada, pp. 131–135.

Jensen, V 2004, 'Measurement of cracks, relative humidity and effects of surface treatment on concrete structures damaged by ASR', *12th Int. Conf. on AAR in Concrete*, Beijing, China. pp. 1245–1253.

Kuzume, K, Matsumoto, S, Minami, T & Miyagawa, T 2004, 'Experimental study on breaking down of steel bars in concrete structures affected by alkali-silica reaction', *12th Int. Conf. on AAR in Concrete*, Beijing, China, pp. 1162–1168.

Larbi, J, Modry, S, Katayama, T, Blight, G & Ballim, Y 2004, 'Guide to diagnosis and appraisal of AAR damage in concrete structures: The RILEM TC 191-ARP approach', *12th Int. Conf. on AAR in Concrete*, Beijing, China, pp. 921–932.

Ludwig, U 1989, 'Effects of environmental conditions on alkali-aggregate reaction and preventative measures', *8th Int. Conf. on AAR in Concrete*. Kyoto, Japan, pp. 503–596.

McCoy, WJ & Caldwell, AG 1951, 'New approach to inhibiting alkali-aggregate expansion', *J. American Concrete Institute*, vol. 47, no. 9, pp. 693–706.

Modry, S 2004, 'Influence of ASR suppressive lithium admixtures on cement paste setting and hardening', *12th Int. Conf. on AAR in Concrete*, Beijing, China, pp. 751–753.

Nilsson, LO 1983, 'Moisture effects on the alkali-silica reaction', *6th Int. Conf. on AAR in concrete*, Copenhagen, Denmark. pp. 201–208.

N.B.R.I. (National Building Research Institute), 1963, *Final report on an investigation of problems concerning the Churchill dam*. CSIR Contract Report CD 228, Pretoria, South Africa.

Nomura, N, Kakio, T, Matsuda, Y & Nishibayashi, S 2004, 'Investigation and repair process of fractured reinforcements due to ASR', *12th Int. Conf. on AAR in Concrete*, Beijing, China, pp. 1271–1276.

Oberholster, RE 1989, 'Alkali-aggregate reaction in South Africa: some recent developments in research', *8th Int. Conf. on AAR in Concrete*, Kyoto, Japan, pp. 77–82.

Olafsson, H 1989, 'AAR problems in Iceland – present state', *8th Int. Conf. on AAR in Concrete*, Kyoto, Japan, pp. 65–70.

Putterill, KE & Oberholster, RE 1985, 'Investigation of different variables that influence the expansion of concrete caused by alkali-aggregate reaction under natural environmental conditions', *CSIR Research Report* BRR 626, CSIR, Pretoria, South Africa.

Rankine, WJM 1862, *A Manual of Civil Engineering*, Griffin & Co. London, U.K.

Reader, CEL & Shaw, JDN 1973, 'Epoxy resin compositions in concrete construction repair', *Symp. on Resin and Concrete*, University of New Castle-on-Tyne. (Available as Shell Chemicals literature.).

Savage, MJ & Cass, A 1984, 'Measurement of water potential using in-situ thermocouple hygrometers', *Advances in Agronomy*, vol. 37. pp. 73–126.

Seto, K, Nishizono, T, Mikata, Y, Maeda, S & Miyagawa, T 2004, 'Maintenance for ASR damaged road viaduct', *12th Int. Conf. on AAR in Concrete*, Beijing, China, pp. 1277–1282.

Silveira, JFA, Degaspare, JC & Cavalcanti, AJCT 1989, 'The opening of expansion joints at the Moxoto powerhouse to counteract the alkali-silica reaction', *8th Int. Conf. on AAR in Concrete*, Kyoto, Japan, pp. 747–751.

South African Institute of Steel Construction, 2010, *South African Steel Construction Handbook (1987, revised 1990 and 2010)*, The Institute, Johannesburg, South Africa, pp. 2.09–2.12.

St. John, DA 1986, 'New Zealand's approach to evaluating the alkali-aggregate problem', *7th Int. Conf. on AAR in Concrete*, Ottawa, Canada, pp. 237–241.

Strauss, PJ & Schnitter, O 1986, 'Rehabilitation of a portland cement concrete pavement cracked by alkali-aggregate reaction', *7th Int. Conf. on AAR in Concrete*, Ottawa, Canada, pp. 210–214.

Torii, K & Kumagai, Y 2000, 'Strengthening method for ASR affected concrete piers using prestressing steel wire', *11th Int. Conf. on AAR in concrete*, Quebec City, Canada, pp. 1225–1232.

Torii, K, Kumagai, Y, Okuda, Y & Sato, K 2000, 'Strengthening method for ASR affected concrete piers using prestressing steel wire', *11th Int. Conf. on AAR in Concrete*, Quebec City, Canada, pp. 1225–1233.

Torii, K, Sanno, C, Kubo, Y & Ohashi, Y 2004, 'Serious damages of ASR affected RC bridge piers and their strengthening techniques', *12th Int. conf. on AAR in Concrete*, Beijing, China, pp. 1283–1288.

Vivian, HE 1950, 'The effect on mortar expansion of amount of available water in mortar', *Studies in cement-Aggregate Reaction XI*, Bulletin 256, CSIRO, Australia.

Wigum, BJ & Thorenfeldt, E 2004, 'Sheets of carbon fibre reinforced polymers (CFRP) as a repair material in order to strengthen and repair concrete damaged by alkali aggregate reactions', *12th Int. Conf. on AAR in Concrete*, Beijing, China, pp. 1289–1298.

Yin, Q & Wen, Z 2004, 'Effects of lithium hydroxide on alkali silica reaction gels', *12th Int. Conf. on AAR in Concrete*, Beijing, China, pp. 801–804.

PLATES

Plate 5.1 One of the test columns referred to in Section 5.1.3.

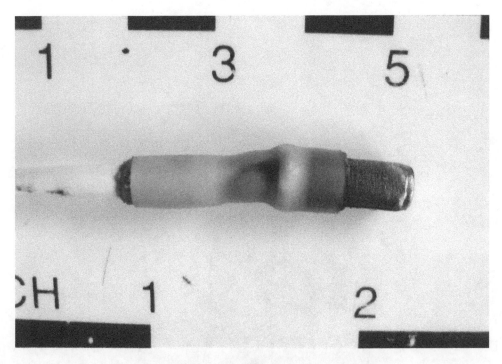

Plate 5.2 Tip of psychrometer. Numbers are cm (above) and inches (below).

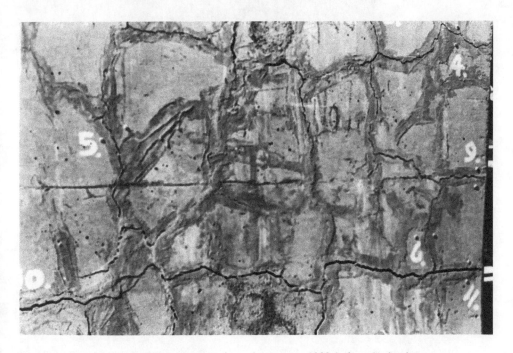

Plate 5.3 Appearance of AAR-damaged surface of column in 1982, before the load test.

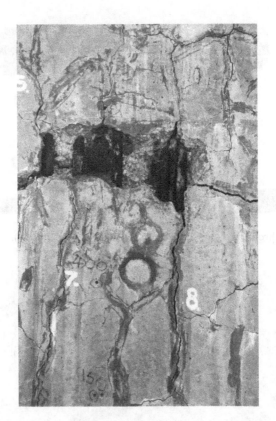

Plate 5.4 Appearance of column surface in 1988. The number 5 painted on the surface in Plate 5.3 is the same as that appearing in Plate 5.4.

Plate 5.5 Hydraulic lift in use to provide access for reading psychrometers.

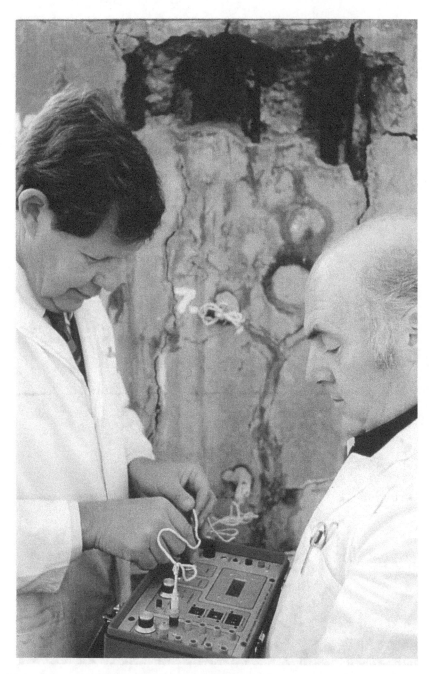

Plate 5.6 Reading psychrometer from hydraulic lift. The cavity behind the observers is that visible in Plate 5.4.

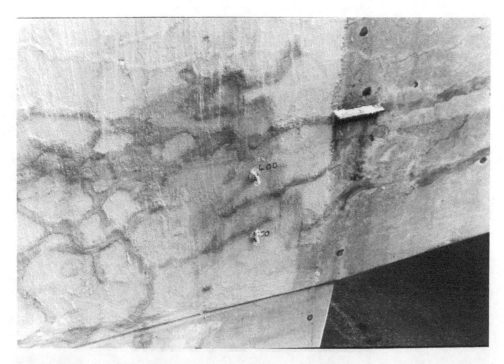

Plate 5.7 Premature treatment of surface of portal beam with "water-proof" coating.

Plate 5.8 End view of sports stadium which is the subject of Section 5.3.2.

Plate 5.9 Deep AAR crack along top surface of cantilever beam.

Plate 5.10 Injection nozzles sealed into surface of beam to carry out resin injection.

Plate 5.11 Appearance of sports stadium columns in 2006.

Plate 5.12 View of end of repaired cantilever projection photographed in 1983. The repaired crack crosses the field of view in line with the halving joint in the beam to the left.

Plate 5.13 Similar view photographed in 2003. The repaired crack is visible near the top of the view and one of the two 1983 coring holes is visible. The anchor plate and one of the tie bars are visible in the centre of view.

Plate 5.14 View taken in 2003 of repair to knee of portal, made in 1979. No work had been done on the repair in the intervening 24 years.

Plate 5.15 Stainless steel torsion reinforcing strips glued on to the spines of twin viaducts damaged by AAR.

Plate 5.16 A stage in the demolition of the portal beam discussed in Section 5.6.3. The main reinforcing was re-used in the re-construction.

Plate 5.17 View of junction between re-constructed beam (to right) and original concrete, photographed 11 years after the repair.

Plate 5.18 An overall view of the reconstructed beam, photographed 12 years after the repair.

Plate 5.19 View taken during construction of a large concrete double-curvature arch dam, illustrating the enormous quantities of concrete used and hence the chance that conditions conducive to AAR might occur.

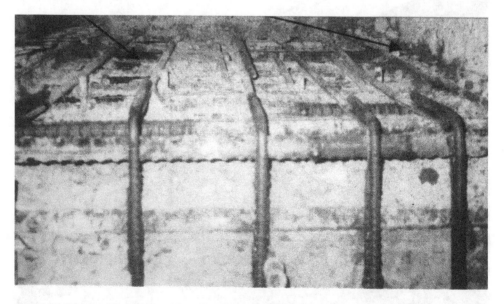

Plate 5.20 Repair of fractured cold-worked high tensile steel bars in a structure damaged by AAR. (Also see Plate 3.1.).

Chapter 6

Epilogue – A check-list of important structural consequences of AAR

6.1 AAR IS A DURABILITY PROBLEM THAT IS UNLIKELY TO CAUSE STRUCTURAL FAILURE

There are no cases on record where AAR has been the primary cause of structural collapse. 'Failure' in this context relates to one or more of the following criteria:

- Unsightly cracking – i.e. aesthetic unacceptability (see Plates 1.1 to 1.5 and 2.1).
- Restrained expansion of a structure leading to secondary stresses arising from misalignment (see Sections 5.8.1 and 5.8.3).
- Deflections that may render the structure unserviceable, or cause secondary damage, e.g. sagging of cantilever ends or long simply supported beams. (This is putative and not proven). Section 5.3.2 shows the reverse, decrease of the deflection of cantilevers as a result of AAR expansion.
- Corrosion of steel reinforcing due to aggressive external agents gaining access to the steel through AAR cracking. Formation of corrosion products may give rise to further cracking of the concrete. (See Plate 5.9, showing an AAR crack that exposed to possible corrosion the prestressing cables of a cantilever).
- The formation of cracks that change the design distribution of loads (see Plates 5.12 and 5.13 showing a situation where halving joints on either side of a cantilever projection were threatened by a horizontal crack in the cantilever).
- Loss of bond between reinforcing and concrete which could result in failure. Laminations formed at reinforcing-to-concrete interfaces have been observed in bridge slabs. An extreme example of a bond failure (however, not caused by AAR) is shown in Plate 3.2.

6.2 AAR RESULTS IN THE DETERIORATION OF CONCRETE PROPERTIES

- The ratio of elastic modulus to compressive strength (E/σ_c) usually lies between 300 and 800.
- Poisson's ratio (ν) is unaffected by compressive strength and the value lies between 0.2 and 0.3.
- The ratio of compressive (σ_c) to indirect tensile strength (σ_{it}) (Brazilian test) lies between 10 and 20.
- The ratio of compressive (σ_c) to direct tensile strength (σ_t) lies between 20 and 60.

All of these ranges were established from the results of laboratory tests on cores taken from AAR-affected structures.

6.3   IN SITU CONCRETE PROPERTIES CAN USUALLY BE EXPECTED TO BE CONSIDERABLY BETTER THAN PROPERTIES MEASURED ON CORES IN A LABORATORY

Theoretical considerations (Section 5.4.3) and laboratory tests (Section 3.5) both show that cracked or even disintegrated concrete retains an appreciable strength as long as it is confined by compressive stresses which could result from applied loading, or from restraint to expansion produced by reinforcing (Section 3.8). This conclusion is borne out by the results of proof loading tests on full-scale in-service structures (Sections 4.10.2 and 4.10.3).

6.4   COMPRESSION MEMBERS ARE RELATIVELY UNAFFECTED BY AAR

As far as compression members are concerned, the effects of AAR on compression members are likely to be minimal or controllable, provided there is sufficient link steel to ensure that cracking is controlled. Concrete in compression acts as a highly redundant material that even in the presence of extensive cracking, can continue to sustain considerable compressive stresses. (See the arch ribs and column in Plates 1.5 and 1.6).

6.5   FLEXURAL MEMBERS NEED MORE CONSIDERATION

As far as flexural members i.e. beams and slabs are concerned, the picture is somewhat different. The most serious consequence of AAR in beams and slabs generally relates to its effect on the bond of steel to concrete. Where bond might be compromised, such as where main reinforcing bars are not restrained by sufficient links or transverse bars, the structural integrity of the member may be compromised. In prestressed members, severe AAR cracking at the ends may compromise end anchorage of tendons.

6.6   THE PERFORMANCE OF STRUCTURAL CONCRETE PAVEMENTS

Performance can be compromised by AAR, but with minimal repair, followed by preventive maintenance the service life need not be reduced. The usual method is to repair cracked joints and slabs followed by a flexible overlay, often of asphalt concrete.

6.7 COMPRESSIVE STRESSES IN AAR-AFFECTED CONCRETE

Stresses induced by the presence of reinforcing, restraint by adjacent structures or friction on the base of a slab on grade, such as a highway pavement or concrete floor can range up to about 4 MPa. In many types of structure, the induced compressive stress may have structurally beneficent effects.

6.8 AAR-DAMAGED STRUCTURES CAN REACH AND EXCEED THEIR DESIGN SERVICE LIFE WITH MINIMAL REPAIR AND PREVENTIVE MAINTENANCE

This has been proven by many examples of AAR-damaged structures that have served out or exceeded their design lives with a minimum of maintenance and repair.

The case histories presented in Chapter 4 show that AAR-damaged structures can meet their design requirements and design lives. The case histories in Chapter 5 show that AAR-damaged structures can be successfully repaired and maintained. It should not be overlooked that all structures, including those not affected by any deteriorative process such as AAR, need continuing maintenance as well as preventive maintenance throughout and beyond their design lives.

Subject index